FIXED MOBILE CONVERGENCE

Voice over Wi-Fi, IMS, UMA/GAN, Femtocells, and Other Enablers

Alex Shneyderman
Alessio Casati

Mc
Graw
Hill

New York   Chicago   San Francisco
Lisbon   London   Madrid   Mexico City
Milan   New Delhi   San Juan   Seoul
Singapore   Sydney   Toronto

The McGraw·Hill Companies

Library of Congress Cataloging-in-Publication Data

Shneyderman, Alex, 1968-
  Fixed mobile convergence : voice over Wi-Fi, IMS, UMA/GAN, femtocells, and other enablers /
  Alex Shneyderman, Alessio Casati.
     p. cm. — (MH technical title)
  Includes bibliographical references and index.
  ISBN 978-0-07-148606-4 (alk. paper)
  1. Internet telephony. 2. Mobile communication systems. 3. Cellular telephone systems.
  4. Telephone. 5. Wireless metropolitan area networks. 6. Telephone systems—Standards.
  7. Convergence (Telecommunication) I. Casati, Alessio, 1970- II. Title.

TK5105.8865.S55 2008
004.69'5—dc22

                                                                          2007049850

McGraw-Hill books are available at special quantity discounts to use as premiums and sales promotions,
or for use in corporate training programs. To contact a representative, please visit the Contact Us pages
at www.mhprofessional.com.

Fixed Mobile Convergence: Voice over Wi-Fi, IMS, UMA/GAN, Femtocells, and Other Enablers

1234567890 FGR FGR 0198

ISBN   978-0-07-148606-4
MHID   0-07-148606-2

Sponsoring Editor Jane K. Brownlow	**Technical Editor** Igor Faynberg	**Composition** International Typesetting and Composition
Editorial Supervisor Janet Walden	**Copy Editor** Bob Campbell	**Illustration** International Typesetting and Composition
Project Manager Arushi Chawla, International Typesetting and Composition	**Proofreader** Laura Bowman	
	Indexer WordCo Indexing	**Art Director, Cover** Jeff Weeks
Acquisitions Coordinator Jennifer Housh	Services, Inc.	**Cover Designer** Pattie Lee
	Production Supervisor George Anderson	

To my family—my wife Olga and sons Alan and Mark—who will always support me through everything. To my parents—"Happy Anniversary, Mom and Dad!" And, to my teachers.

—Alex Shneyderman

I would like to dedicate this book to all my friends and loved ones, who have suffered (along with myself!) from my retirement from normal life during the writing of this book, but who also have encouraged me in the process.

—Alessio Casati

About the Authors

Alex Shneyderman has held a variety of engineering, technical marketing, and business development roles with both startups, such as SpringTide Networks and Dynamicsoft, and major corporations such as Alcatel-Lucent during his 12 years in the telecommunications industry. Currently at Motorola, Alex is taking a key role in positioning advanced mobile technology with leading North American telecommunication service providers as a technical business development manager. He has authored a number of papers, tutorials, and other publications, including *Mobile VPN: Delivering Advanced Services in Next Generation Wireless Systems* (Wiley, 2002), a book he authored with Alessio Casati in 2002. Alex's current interests include aspects of convergence as well as machine-to-machine and remote sensing technology.

Alessio Casati is currently the technical manager of 3GPP Core network standards at Alcatel-Lucent. He has been active in the industry of mobile and data communication for the past 12 years in research and development, standardization, as well as product development and solutions engineering roles. He holds several patents in the field of mobile data communications and has authored numerous publications, including *Mobile VPN,* along with Alex Shneyderman, coauthor of this book. He is currently focusing his efforts in the areas of the evolution of 3GPP mobile systems, IMS, and convergence.

About the Technical Editor

Igor Faynberg is the technical manager of the Next Generation Network and Internet Standards group in the Convergence CTO Solutions, Architecture, and Standards organization of Alcatel-Lucent. He is also an adjunct professor of computer science at Stevens Institute of Technology.

Contents

Foreword

It would be an understatement to say that a book on fixed-mobile convergence is very timely. Major telecommunications corporations—both service providers and vendors—have created, this very year, the namesake business groups. Furthermore, the underlying technologies and standards are being embraced not only by the cable and IT industries, but also by those of media and advertising. Sure enough, this multi-industry attention, which employs a huge part of the world population, can be hardly ignored by the regulators, economists, and academics worldwide. Many people need to catch up, to learn, and to plan for the future in this frantic environment—and what better source of learning than to read a book on the subject?

And it is a very complex subject at that! There are two major aspects to fixed-mobile convergence. First is the *technology* aspect, which encompasses everything from signal processing and radio interfaces, to the inner workings of the present public telephony, to the nuts and bolts of the Internet Protocol and its numerous applications. Second is the business aspect: The impetus of convergence *will* make us much better connected both to one another and to the information we need. That will, in turn, make us much more efficient at what we do while working, shopping, banking, and learning from home—or, for that matter, any other place we choose. That increase in efficiency, pervading virtually all fields of work, naturally translates into an unprecedented business potential. Consequently, we should expect some start-ups to grow into near-monopolies, threatening the incumbents; we should expect some basic communications services to become much cheaper than they are today (or even free); and we should expect many new services to be available to us. Both aspects are very much interrelated in that it is impossible to have a *business vision* without thoroughly understanding the underlying technologies, and it is impossible to envision the convergence of a plethora of technologies without understanding the business. The crux of this interrelation lies in the technical standards that reflect the latest research in the industry, as it makes the ultimate decisions on implementing the technical substance of its business vision.

This book addresses both aspects, and it addresses them remarkably well. There is a good reason for that: The authors Alessio Casati and Alex Shneyderman, who belong to the Internet generation, have spent their formative working years at Bell Labs, where most of the involved technologies were developed and where much of what is in the news now had previously been written on the whiteboards and discussed in the hallways for years. Both authors (for whom this is the second book) have influenced the development of the technologies they describe and distinguished themselves in business as well. Alessio Casati has been involved with engineering mobility products while contributing to the key convergence standards in the Internet Engineering Task Force (IETF, the Internet think tank) and the Third-Generation Partnership Project (3GPP). Currently, he manages 3GPP Core Network and System Architecture standards as part of the CTO office of the Convergence Business Group of Alcatel Lucent. Alex Shneyderman has been excelling in both engineering and product management and marketing while gaining unique business experience and contributing to the success of three start-ups before taking the technology business development position in Motorola.

I am hardly new to the field, but there was still much for me to learn from the manuscript—especially in the area of business development. I have no doubt that engineers, businessmen, lawmakers, and students—all who want to further contribute to or stand to benefit from fixed-mobile convergence—will find the information they need in this book.

—Igor Faynberg
Technical Manager,
Alcatel Lucent

Acknowledgments

The authors would like to acknowledge significant contributions and help from numerous individuals and organizations. This book would not be possible without generous support from Igor Faynberg, who not only introduced us to McGraw-Hill but also tech edited the book, providing invaluable guidance and deep insights in various areas of technology covered in the book.

We would also like to thank our colleagues at Alcatel-Lucent, Motorola, and other companies for valuable discussion and their help and contributions to both technical and business-related materials covered in the book.

This book would not be possible without the McGraw-Hill team's diligent management and guidance during writing and publishing. Our special thanks go to Jennifer Housh, Arushi Chawla, and Jane Brownlow, who were instrumental in bringing this book to life. Finally we owe deep gratitude to our immediate families, Alex's wife Olga and two sons Mark and Alan, and Alessio's friends and loved ones for their patience, support, and encouragement during the difficult writing process.

Introduction

Consumers and Their Discontent

A few years back, talking to a dentist friend of ours, we were shocked to find out that he was convinced that mobile phones work off the satellites up in the sky—how an educated person in our enlightened times can be so technically unaware! We then kept him amused for about 15 minutes with a description of a highly developed terrestrial infrastructure of cell sites, access network, backhaul transport network, core network, and so on. That small incident, however, got us thinking: What does user experience have to do with the underlying technology? People have other things to worry about when selecting a service or product. It has to fill a set of needs, be easy to use, and have a reasonable price—as any MBA student should readily confirm.

Take telecommunication, for example: Why should today's subscriber to telecommunication services be constantly aware of the technology being used and juggle between different and often very dissimilar devices, phone numbers, and bills? Why should you, the consumer, have to search for a mobile phone to look up a contact in a contact list to make a call on landline? Why are up to 20 percent of all SMS messages sent to landline numbers (in countries with a common numbering prefix for landline and mobile numbers) and lost forever? Why are up to 10 percent of all enterprise and institutional users keeping their landline phones forwarded to their mobile phones 50 percent of the time, and as a result often paying double fees?[1]

Pity the poor user! Individuals and organizations paying a premium for fixed and mobile telecommunication services do not have to be content with a thus-degraded user experience or an antiquated landline communication model dating back more then 100 years. The users should not be forced to rely on their technology savvy to make the most of the service or product. And that is just the beginning of our problems. There is a greater hurdle (stemming from the technology gone rampant) to overcome: complexity.

---

[1] ABI, Gartner, and others.

A Specter of Complexity

A specter is haunting the telecommunications industry—the specter of *complexity*.[2] The industry as we know it may be on the verge of a crisis: the crisis of complexity, that is. Multiple types of network access—voice and data, fixed and mobile, wireless and wireline—present the user with an increasingly difficult-to-manage array of communications options. Add to this picture new communications methods—instant messaging, push-to-talk, SMS, content sharing—and multiple devices needed to make use of all of this technology—laptops, traditional and IP phones, Blackberries, cellular phones, smart phones, and connected PDAs and media players—and it becomes easy to appreciate the degree of user frustration.

Imagine now a seamlessly connected world—like in a science-fiction novel—a world in which telecommunications is replaced with communications and the "tele" no longer has any significance. A world in which services are converged, the access network type no longer matters, and your communications experience is simplified with a single "any-media" service and devices with intuitive user-centric UI and functions. Now mix in some never-before-possible solutions such as unified presence and messaging or location and content-aware applications, and the outcome becomes very predictable.

This vision may allow the industry in perpetual search of a next killer application (killer app)[3] and device form factor to reinvent itself and truly become experience-driven and user-focused. This, in turn, may eventually aid the telecommunications industry, which is entering its maturity, to instead continue the growth phase, enabling new opportunities.

The good news is that the solutions making this vision a reality already exist under our fingertips. In fact, a brand-new field known as fixed-mobile convergence (FMC) is being created as you read this book . . . or are about to! FMC is based on a variety of technologies and standards centered around the Internet Protocol such as the IP Multimedia Subsystem (IMS), Unlicensed Mobile Access (UMA), femtocells, Voice over Wi-Fi (VoWi-Fi), and others. While only recently developed and defined, this field is benefiting from the experience gained through trials and errors in delivering services to consumers, and prides itself on being born to address specific user and service provider needs.

What This Book Is About

The book accomplishes a number of important tasks. First and foremost, it discusses how the telecom industry is reshaping itself and moving toward convergence of fixed and mobile technologies, enabling a uniform seamless telecommunications experience. It provides a perspective of a business case and value proposition of fixed-mobile

---

[2] Paraphrasing an infamous quote from Friedrich Engels' and Karl Marx's *The Communist Manifesto*: "A specter is haunting Europe—the specter of communism."

[3] A snappy term first used by Mitch Kapor, the founder of Lotus Development corporation, in the mid-1980s to describe Lotus 1-2-3 once it became evident that demand for that product had been the major driver of the early business market for IBM PCs.

convergence in both residential and enterprise markets. It also analyzes the fallacies and pitfalls of convergence in real-life deployments, and addresses common confusions and misconceptions about technology and definitions in the field.

Second, the book serves as a comprehensive reference text on technology of the newly created FMC field, its variants, and its main underlying components, such as various convergence mechanisms, IMS, Voice over Wi-Fi, femtocells, and other relevant technologies such as WiMax, presence, location-based services (LBS), push-to-talk (PTT), machine-to-machine (M-to-M), and universal instant messaging (IM).

Finally, the book presents an in-depth analysis of FMC solution architectures and practical implementations, targeting both the vendor and the service provider communities. Special attention is paid to the analysis of real-life user experience scenarios and the business impact of FMC solutions and technology.

The Structure of This Book

A clear and consistent structure is a must for a decent technical book. It helps the reader to navigate the subject effectively and makes the book an invaluable reference tool, guaranteeing it a long service life on an eye-level shelf. That's why we spent a good amount of time tuning this book's structure.

After a brief excursion in the world of business and a close look at the past and present convergence drivers and deployment experience, we turn our attention to an in-depth discussion of the main subject of the book: defining convergence fundamentals. We are dealing with the complexity of the subject matter and its somewhat loose nature by essentially decomposing the nascent field of FMC into its natural components: *fixed, mobile,* and *convergence.*

The "Fixed" section builds a necessary knowledge foundation and provides an adequate description of the underlying technology. Here we give the reader a tour of the legacy signaling and media technology and a brief overview of the infrastructure making up today's traditional plain old telephone service (PSTN) and Voice over IP (VoIP).

The "Mobile" section deals respectively with mobility technologies. We start with a deep dive into cellular and WiMax technologies, and add Wi-Fi to the mix. The section concludes with a discussion of the real-life deployment issues and a gradual transition to the topic of convergence.

The "Convergence" section discusses technologies and a service framework enabling convergence between fixed and mobile voice communications. The discussion in the chapter is not limited to technologies of the day such as UMA, IMS, VCC, and femtocells. Here we are taking a more fundamental, technology-independent approach focused on the user experience and objectively evaluating a number of competing, sometimes unexpected technologies enabling the experience of seamless connectedness.

The book concludes with chapters analyzing relevant use-case scenarios and specific experience enablers as well as advanced services and applications affected or enabled by convergence, taking a closer look at the standardization effort in the field and finally looking ahead to the future of FMC and its impact on the way telecommunications will evolve in the coming decade.

You, the Reader

One of the challenges faced by the authors in the writing of this book was one of exclusion vs. inclusion. We were careful to cover only the relevant aspects of the established and well-described technologies composing the FMC and at the same time to provide an in-depth discussion of the new fields, especially the ones not extensively covered in the literature. Consequently, this book requires you, the reader, to have a reasonable level of understanding of fundamental concepts of fixed and mobile communications.

Our other challenge was to make this book useful for as diverse an audience as possible by writing interesting text not only for professionals in the fields related to convergence and telecommunications in general, but also for the readership from adjacent fields, such as investors in the sector, analysts, and—we hope—casual readers interested in one of the most promising new directions of modern communications.

Let's Converge Fixed and Mobile!

So here we go again. The industry has tried everything at its disposal to converge fixed and mobile communications, from pricing-based strategies and simple forwarding of calls, to elaborate network-wide, multiservice, IN-based solutions. Now let's ask ourselves, do we really believe this latest spin on yet another attempt to converge fixed and mobile will finally make it?

Clearly, if we did not think so, we would not be writing this book. The massive changes in telecommunications landscape induced by the transition to access-independent, IP-based communications is one thing that is different this time around. We also believe that growing consumer discontent with current disparate, overly complex telecommunications service models and their desire to *simplify* has reached the tipping point and will pressure operators and vendors to come up with viable FMC solutions. Indeed, if convergence does not start happening now, the industry will face imminent stagnation and eventual decline.

So let's delve into the details and let's have smooth sailing. We have striven to make this journey a useful and interesting one for you.

Introduction to Convergence

Convergence is the property or manner of approaching a limit, such as a point, line, function, or value.

—Encyclopædia Britannica

Convergence is a property of organisms not closely related, that independently acquire similar characteristics while evolving in separate and sometimes varying ecosystems.

—American Heritage Dictionary

Convergence is similarity of form or structure caused by environment rather than heredity.

—Webster's Dictionary

As you would expect at the beginning of a book such as this one, the first chapter provides the initial introduction to the topics that are discussed throughout the text. It also builds the foundation for more in-depth discussion on fundamental technology and business matters to follow. Here we are defining the terminology and concepts as well as setting the scene by describing the telecommunications industry, where global trends such as "convergence," "migration," and "substitution" are taking place. Of all these aspects, "convergence" in particular is a necessary step for the industry players to ensure that they have a place in *the future* of telecommunications, which is becoming more and more its *present*.

What Is Convergence?

It is a good idea to start the discussion on the topics about convergence of *fixed* and *mobile* communications, which is now heavily influencing the telecom industry with a precise working definition of the term "convergence." Definitions are something we are

Transition of two
dissimilar entities to
become similar

Similarity of form or structure
caused by environment
rather than heredity

Convergence

Manner of
approaching a limit

Independently acquiring
similar characteristics in
different environments

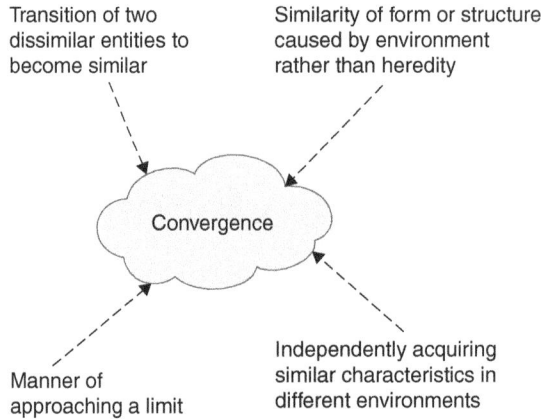

Figure 1.1   Convergence defined

going to be giving special attention to throughout the book, perhaps, not with the ambition to create yet another telecommunication dictionary,[1] but rather to create a solid basis from which to explore the subject. The definitions we are going to provide mainly have significance in the specific area of technology; thus our usage of the terms will be limited to the specific field of telecommunications.

There is no shortage of opinions within the industry when it comes to defining the meaning of the term *converged communications* or, as it is now often called, *seamless mobility* (see Figure 1.1). Any definition is bound to have some degree of imprecision, as convergence of telecommunication services can also apply to users that are not moving at all! In fact, depending on the context, the audience, and the speaker or author, the terms often take different, sometimes unexpected and even opposite, meanings.

In its general form, convergence is an act of a transition of two or more dissimilar entities to becoming similar. Applying this definition to telecommunications, we further define convergence as a trend where distinct communications systems, services, or devices evolve to provide users with the same or a similar experience. It is important to note that the components of telecommunications systems themselves do not have to be the same or similar—that is, as long as they make a uniform user experience possible.

The specific form of convergence we are interested in can be very well defined or at least characterized by any of the definitions quoted in the epigraph. In fact, many in the industry these days view convergence as a next big step for telecommunications, the ultimate connectivity state when all networks come together to create an array of seamless, always-connected services across them. Further, even though modern telecommunications technologies are often not "closely related," the need to enable uniform user experiences and the pressures of competitive environment force them to "acquire similar characteristics while evolving in different domains."

---

[1] Not that we are incapable, but rather our readers may not appreciate the effort.

What Is Being Converged?

As we have the convergence definition out of the way, now let's take the next step and look at a specific type of convergence, the fixed-mobile convergence (FMC). To facilitate the discussion, we are going to apply the proven approach of decomposing the technology and services into distinct layers or stages.

Before we can start drilling down to details, however, we need to clarify the important distinction that exists between familiar but often misused terms: fixed, mobile, wireless, and wireline. This distinction, while clear to most professionals, remains a bit vague for many outside the industry. Indeed, why isn't fixed-mobile convergence not called wireless-wireline convergence? Let's find out.

Wireless is not synonymous with mobile,[2] just as landline or wireline is not the same as fixed. Often the terms are used interchangeably; however, they actually mean different things, especially in telecommunications, so the fixed-mobile convergence is not called wireless-wireline convergence, and not only for the reason that the acronym would sound somewhat awkward, something like WWC.

Let's first analyze the difference between fixed and wireline. In telecommunications, the term *wireline* (or landline, as it is sometimes called) refers to a type of communications system requiring the propagation of a signal over solid media such as copper wires (like those used to connect fixed telephones to the telephone network), glass fibers, or coaxial cables (such as those used to connect a TV set to the cable network) to transmit information between two endpoints. *Fixed,* on the other hand, refers to a method of communication that requires the user of the fixed communications service to maintain its physical location relatively constant to be able to communicate remotely.

As you can see, the distinction is obvious. Indeed, fixed communication service is often provided over wireless media, for example, point-to-point microwave links, optical last-mile solutions, line-of-sight laser communication, the fixed flavor of WiMax (defined by the IEEE standard known as 802.16 [114]), and other "last mile" technologies. Even wireless LAN access available in hotel rooms or in airports or train stations, which most people have come to experience one day or another in their lives, can be considered a fixed service in most cases.

The distinction between wireless and mobile is less clear; however, it is more important for the discussion ahead. *Wireless* in the telecommunications industry usually means the type of communication that is conducted over untethered media such as air for cellular communication or vacuum in space for satellites. Mobile communications, however, assume it is possible for a mobile user to maintain uninterrupted connectivity to a network or a communication peer, or continue using a telecommunication service while changing location, or method of connectivity over time. A mobile user would stay connected and oblivious to changing conditions as if these events were not happening and that user were not moving at all (that is, ultimately, as if the user were using a fixed communication service).

---

[2] In all fairness we have to note that despite a clear distinction in theory, in practice, mobile services have almost always been provided over wireless networks.

Systems capable of mobility support are therefore called *mobile systems,* and the end-user devices supporting mobility are called *mobiles, mobile stations,* and *mobile nodes.* When entire communication networks are on the move and use a mobile communication system, these networks are also known as *mobile networks*[3] (often used in military communications applications, the cruise ship industry, and space exploration).

Like fixed networks, mobile networks can be implemented over both wired and wireless media, particularly with the help of FMC. With most FMC solutions the user can seamlessly switch between different fixed and mobile communications methods while remaining connected, as fits the definition of mobile service. In the context of FMC, mobility can be further decomposed into terminal and service mobility. *Terminal* mobility is defined as the ability of a subscriber or service to use the same terminal equipment or mobile device for different types of communication. *Service* mobility, on the other hand, refers to the user's ability to access the same set of services regardless of geographical location or method of communication.

In summary, based on this discussion, it is then clear that "fixed" and "mobile" are service attributes, while wireline and wireless are technology attributes. Both wireline and wireless *technologies* can be used to deliver fixed or mobile *services.* For this very reason the term was coined as fixed-mobile convergence, as this is a convergence of telecommunication services, rather than a convergence of technologies, which will remain distinct!

Convergence or Substitution?

Having defined and analyzed at a high level the *F* and *M* components of FMC, let's see how the *C,* that is, convergence, measures up against another important trend in modern communications, substitution (or migration, as it is sometimes referred to in the industry). Indeed, many in the field are convinced that substitution, not convergence, will take center stage in the industry over the next decade. After all, who needs fixed services when mobile-only solutions can take care of many of the user needs and wants FMC is called upon to address? To answer this question we will need to dig deeper into what substitution really gives us.

Unlike convergence, substitution does not require extensive definitions. It is a simple shift of the bulk of usage from one communications mechanism to another, for example, from telegraph to telephone or from fixed to mobile. The latter recently progressed from a barely visible trend to a small revolution, especially in some regions of the world such as Western Europe and parts of Asia. Many users in these geographic areas are either completely replacing their fixed-line service with mobile or shifting the bulk of their connectivity needs to mobile, which dramatically reduces fixed network usage.

Not surprisingly, this shift is mainly applicable to voice communications. While mobile data is definitely on the rise (up by 46 percent in 2005 alone as depicted in Figure 1.2), its usage (and capability) is currently nowhere near that of broadband technology such

---

[3] Note that the term mobile networks is also widely used to refer to any networks providing mobile services.

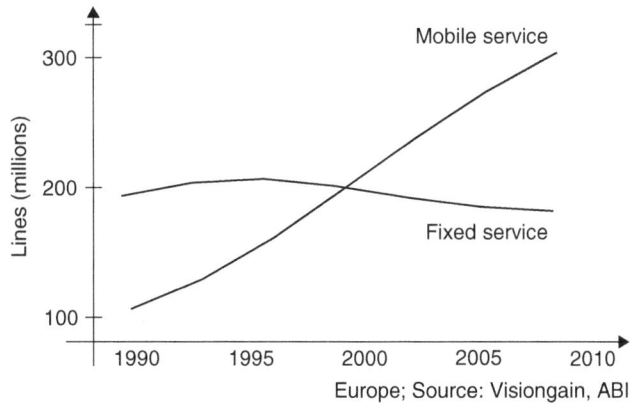

Figure 1.2 Fixed-to-mobile substitution on the rise

as DSL or cable. The decision to use either fixed or mobile services often represents a trade-off between cost, functionality, and convenience. Almost always users have to decide whether to exploit the convenience of mobile anywhere/anytime reachability or to forego this more expensive technology and settle for a cheaper[4] yet often more reliable fixed option. Data services, which generally can offer sustained bit rates one order of magnitude higher in fixed than mobile environments, offer a great example. In fact, high-speed data access in residential environments today and in the foreseeable future will almost always require a fixed line.

Substitution Barriers

There are many reasons for the rise of substitution. Despite the typically higher price of mobile services, mobile devices' and services' biggest advantages are their higher geographical availability and vastly richer feature set. These attributes dramatically increase perceived service usefulness and appeal to the user while offering a far greater potential for mobile calls to be placed than fixed ones.

According to ABI research, fixed-to-mobile migration often does not depend on geography but instead is tied to certain consumer segments such as single urban professionals or students who increasingly prefer to ditch their fixed voice service altogether in favor of mobile-only communication. This trend is also helped by ubiquitous mobile coverage, falling mobile minutes prices, and frequent promotions such as free night and weekend calling and equipment subsidies. So why then do we believe that convergence, especially FMC, will be one of the major driving forces in telecommunications of tomorrow? The answer to this question has to do with the fundamental properties of both fixed and mobile communications as well as with specific user experiences enabled by

---

[4] In some cases this may not even be cheaper.

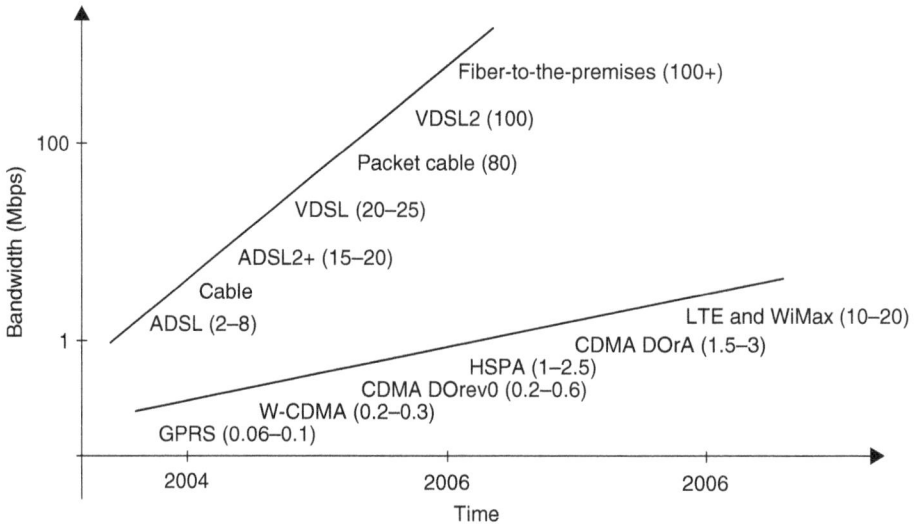

Figure 1.3 The need for FMC: bandwidth

corresponding offerings. The fixed communication service properties that constitute barriers to substitution include:

- Bandwidth (in foreseeable future) and Quality of Service (QoS)
- Cost
- Coverage (in certain areas)
- Security
- Service composition richness
- Equipment convenience (in most cases)

First and foremost, fixed communication will be one step ahead of mobile communication in terms of sustained bandwidth and service cost for a long time to come, therefore making it more attractive to many enterprise and residential user segments for which ubiquitous mobility may not be such an immediate priority to forego the benefit of high data rates of fixed services (Figure 1.3 further illustrates the trend).

Speaking of which, mobile coverage is not expected to become uniformly[5] ubiquitous in our lifetime. Indeed, in many (most) areas of the world, fixed-line communication is still more pervasive than wireless coverage, and will remain the only reliable means of communication until significant investments are made in improving cellular infrastructure (see Figure 1.4 for a map of typical U.S. mobile coverage).

---

[5] That is, except for certain densely populated geographical clusters such as Europe, parts of the U.S. East Coast, and some East Asian countries.

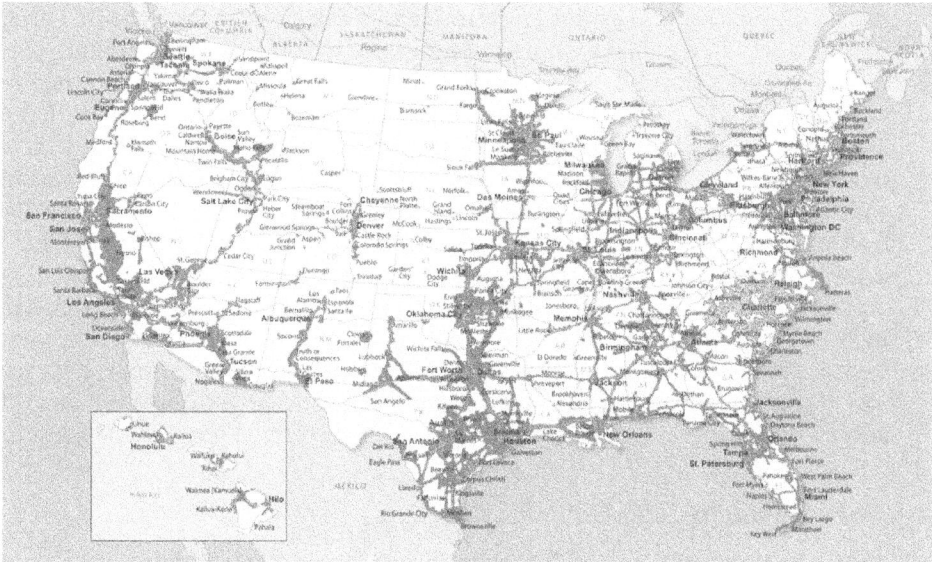

T-Mobile U.S. coverage map

Figure 1.4   The need for FMC: coverage

Conversely, in many developing countries, wireless communication is the prime way to stay connected, as the necessary investments to develop a ubiquitous wiring of premises simply were not possible in the past and still lack adequate backing. But in these countries we cannot then technically call this wireless substitution, as there is simply no sufficiently developed wireline network to be substituted!

Another reason is security. A wireless mobile network will always be more exposed than a wired network in terms of security. This is in no small part due to a simple fact that breaking into a conversation or data session conducted over a fixed line requires physical tapping into the wires—a step unnecessary for successful wireless snooping. This limitation of course crosses the border of the two realms when a wireless medium is used to access fixed communication services (for example, when a wireless router is used at home to access a DSL broadband network service). In this case, setting up security is a user choice and responsibility, and the environment and expected usage may determine the user's security sensitivity.

As this phenomenon is more and more recognized, developers of applications used both over wireline and wireless access in mobile and fixed environments do not normally rely on Link- or Physical-layer access security; instead, they have developed Network- and Application-layer security models (similarly to VPN used for remote access to corporate LANs, or SSL used to access online banking or trading sites, or even VoIP applications such as Skype).

One more reason is that wireline networks are increasingly used to deliver a wide variety of fixed services (such as IPTV and content sharing), thanks to steadily increasing data rates due to fast-paced evolution of access technologies, for instance DSL, cable, and fiber to the curb (FTTC). This allows combining the delivery of a multitude of data, video, and speech applications at a price point not yet possible in mobile-wireless environments.

Finally, fixed terminals and customer-premises equipment (CPE) will for a long time be a step ahead of mobile devices in terms of cost, convenience (large, comfortable-to-hold handsets, large buttons and displays, longer battery life, internal communication capabilities such as PBX phones in enterprises, or multiple bases and handsets for cordless communication in the house), and ease of use if not feature content and service. That will continue to give them high appeal in the eyes of many subscribers.

Even when operators decide to provide service that is perceived by users as strictly mobile, they may be tempted to resort to providing parts of this service over fixed lines for the aforementioned reasons. That's when FMC comes in to make coexistence of fixed and mobile smooth, natural, and above all *imperceptible* to the end user. The resulting service typically combines the advantages of both fixed and mobile communications.

From Convergence to FMC

As we have established in the preceding section, both fixed and mobile telecommunication services will remain available for many years to come. For this very reason we will see emergence of FMC solutions in many possible forms with its main goal being to blend fixed and mobile methods of communication, providing the user with uninterrupted, location- and network-independent access to the same service set. The user of an FMC solution is typically offered the same or a similar experience regardless of the type of communication system in use at the time and is not aware of changes in communications media or any other properties of a system in use at any given moment. While the primary focus of FMC is convergence of voice communications, more and more attention is being paid to data and applications other than voice.

FMC Foundation

Fixed-mobile convergence can be based on a variety of technologies. Regardless of the underlying mechanism, however, most FMC solutions share a common foundation or model. This model can be best defined in terms of convergence of four layers—Transport (aka Network), Service, Application, and Terminal—as depicted in Figure 1.5.

In the spirit of FMC (brought to life by desire to simplify and improve the existing telecommunications experience rather than to create yet another technology), the model is defined mainly in terms of user impact.

Transport Convergence   As evident from its name, the FMC *Transport* layer is primarily responsible for making convergence possible in the core. The Transport layer may, for example, include technology allowing for seamless switchover or interworking of voice traffic between different kinds of networks such as IP-based wireline networks and wireline circuit–based ones, or enabling smooth migration of a session in progress between different types of wireless networks such as cellular and Wi-Fi.

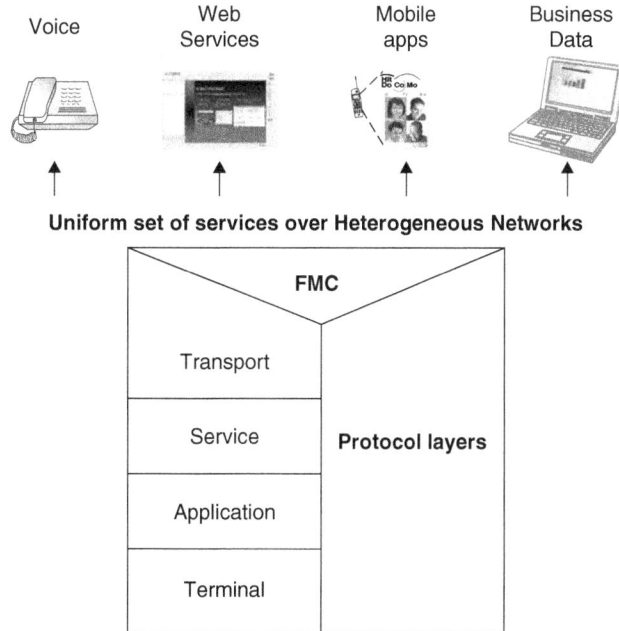

Figure 1.5 Four layers of FMC

As such, the technology implemented at this layer is responsible for interoperability, handover, and integration of aspects such as the security and operation, administration, management, and provisioning (OAM&P) components of different networks playing a role in the FMC solution—not only heterogeneous, like fixed and mobile, but also homogeneous, like dissimilar kinds of mobile networks. In addition, this layer typically provides routing and switching of data and packet or circuit voice, along with media and signaling control and integration.

Service Convergence   The Service layer of the four-layer model provides support for service decomposition into building blocks that enable a single set of services to be offered over multiple heterogeneous networks. This layer enables a common service framework independent from the underlying transport network to be used for the majority of the existing end-user applications. Moreover, this layer enables new, previously impossible, applications such as shared presence, media-independent videoconferencing, collaborative gaming, and any-media instant communications. The Service layer is also responsible for session continuity while switching between different networks and creation of a common shared user profile and presence state repository that can be shared by multiple applications.

To achieve these goals, this layer relies on specific Transport layer technologies such as Mobile IP or General Packet Radio Service (GPRS) for data mobility and Universal Mobile Access (UMA/GAN [80], [81], [82]) or Voice Call Continuity (VCC [76]) for voice call continuity across wireless accesses of different types of wireless access (i.e., circuit and packet based).

The other key feature of the Service Convergence layer is active "session awareness" and the ability to implement specific policy actions, regardless of the location of session endpoints; session awareness and related policy services are fundamental building blocks in delivering converged services. They are key enablers of seamless roaming of data and voice sessions across wireline and wireless broadband domains; the network will dynamically adapt its policies to cope with the transitions of terminals from one medium to another.

Application Convergence   The Application layer is what subscribers actually interact with when using FMC services. The idea behind this layer is to make the experience of switching between different means of communication transparent for the user engaged in a session.

That is, the FMC service provider's goal is to provide a consistent user experience independent of the type of device or network used, so the applications themselves might be called upon to round up the FMC solution if the perfect transition between different types of media or devices has not been achieved at the Transport or Service layer. Examples of such applications may include push-to-talk (PTT) service between different endpoints such as a phone and a desktop PC, voice conversations continuing without interruption in the office via Wi-Fi and PBX or over wide area cellular networks, and device- and network-independent presence-enhanced contact lists. Functions typical of this layer may also encompass codec adaptation and content conditioning. Applications helping to enable convergence do take an important role in the FMC ecosystem. Over time they will make new and unique user experiences possible and will ensure the eventual success of FMC by targeting specific user segments, and by offering new and unique service bundles.

Terminal Convergence   The Terminal Convergence layer is principally concerned with the end-user terminal and its ability to support converged applications and services. Device manufacturers are typically trying to solve both hardware and software issues common for these devices by incorporating features of both fixed and mobile terminals.

And here lies the dilemma: The properties that make mobile phones attractive, like small size, ruggedness, and suitability for outdoor use, are often either irrelevant to the landline devices user experience or in direct contradiction with their common properties such as large screens and buttons, industrial design (ID) optimized for comfort during long conversations, and cradle charging. Table 1.1 summarizes the major advantages of each class of device.

Perhaps one way to resolve this dilemma is to create families of multimode devices geared toward specific experiences and usage patterns while sharing a common UI and aspects of industrial design. For example, the user may be offered a dual-mode device similar in form factor to today's mobile phones for use on the road and occasionally indoors, and another dual-mode device implemented as a cordless phone or tablet PC for use in fixed locations such as an office or home and occasionally in wide areas.

Therefore, the main challenge to overcome for the technology at this layer is going to be resolving these contradictions by creating new multifunction terminals with unique

TABLE 1.1   Mobile vs. Fixed Devices

Mobile	Fixed
Optimized for outdoor use	Optimized for better ergonomics
Optimized for longer battery life	Optimized for longer conversations
Mobile form factor	Inexpensive
Wide-area radio	Tethered or short-range radio

Feature-rich; often multifunction:
   PDA
   Music player
   Video
   Camera
   Navigation
	Basic functions:
	   Voice
	   Voice mail
	   Contact list

form factors and properties, and creatively adapting the existing devices for the FMC use-case scenarios and experiences.

In addition to form factor and UI issues, the composition of features the radio- (or even fixed line termination–) converged terminals should support is also a critical success factor for the delivery of an appealing set of FMC services. The combination of these technologies in a manner that is not predefined is one of the justifications of the layering approach that attempts to decouple the access technology–specific aspects from the FMC user experience.

Important Caveat   No model is perfect or fully inclusive. Not all layers of this FMC model are present in any FMC solution. The model we have introduced is more of a framework, a set of guidelines around which one can build a successful service or product. Thus, the notion of a Transport layer is applicable only to FMC solutions relying on network convergence. The solutions based entirely on terminal convergence or application convergence may not be concerned at all with the type of underlying network.

Yet another group of solutions implemented entirely in the network (such as femtocells—more on those in Chapter 5) may not require any application or terminal convergence at all. Albeit not providing a fully converged experience, these solutions nevertheless deliver sizable improvements over the existing disparate networks, devices, and services.

Brief History of Fixed-Mobile Convergence

While certainly on the rise in later years, fixed-mobile convergence is not a new technology. In fact, it has been around in some form since the first mobile calls were made. More precisely, the industry and often the end users themselves have *attempted* to converge different types of communications solutions. To date these attempts have not been very successful. It was just hard to swallow for both operators and users alike that different modes of communication will stay vastly different.

So the industry tried and tried. Each time, everyone thought the new technology would save the day—but it never did. Over time, through trials and error, the industry has come to realize that the success of a service offering is in the user experience, a compelling business case, and satisfying specific needs of users and service providers—not in the technology. Thus the term fixed-mobile convergence was coined and the field got its early start.

We ourselves witnessed a few such initiatives, all of which either failed in the market, are still in development, or are in the midst of increasingly long sales cycles. There were also a handful of attempts to enable convergence in some form through creative pricing plans (including home zone pricing in Singapore, the UK, Germany, Italy, and the U.S.).

Initially the big seller to operators was that FMC was going to be the silver bullet to the retention issue, even if it cut the operator's revenue per subscriber. But that need was addressed through a simpler solution based on subsidized service contracts, unique user interfaces, and distinctive features. That's why we see these same marketing techniques being adopted by landline operators increasingly offering contracts and bundles. In their service offering, wireless operators try to completely eliminate the need of a fixed-line phone. Wireline operators in turn try to act as mobile virtual network operators (MVNOs) and set up partnerships with some wireless carrier to get access to a common customer base.

From a service standpoint, wireless single-rate ("all-you-can-eat" for a fixed fee) plans combined with free in-network calling, free nights and weekends, and other perks, coupled with fast maturing technology, decreased the need for FMC during the last two decades.

Table 1.2 shows some of the past FMC attempts and ranks them in terms of commercial success and user experience and usability. It also facilitates the discussion that follows, providing a short summary of the past technologies and briefly analyzing the reasons for their failure or relative success.

TABLE 1.2   Past FMC Attempts

FMC Solution	Drawbacks
DECT/GSM	Voice only, separate numbers, no true integration, expensive AP/base station, manual switchover
Wireless PBX	Complex implementation and integration, limited functionality, lack of true integration
Cellular hotspots	Voice only, lack of service transparency, lack of home number and cellular integration
Auto-forwarding	Voice only, network utilization, lack of functionality, nonintuitive, no true integration, only works for incoming calls
Billing/tariff-based solutions	Addressing only few aspects of a larger problem, failure to offer convergence, limited usability
IN-based solutions	Hard to implement, costly

DECT/GSM

Digital Enhanced Cordless Communication Telecommunications (DECT) is a cordless technology that was originally designed to be complementary to GSM on a landline in a high-density office environment. Table 1.3 compares the two.

GSM is a wide-area technology with a proportionately higher cost per base station (compensated by the need for fewer base stations than DECT) and potentially low-priced handsets. DECT, by contrast, is better suited to high-user-density, small-area coverage (actually, its current use is essentially limited to home and office environments), with correspondingly lower base-station costs.

The FMC services based on DECT are strictly dual-mode handset-based and do not require integration of mobile and fixed services in the core. DECT combined with GSM is essentially a smart-forwarding solution directing user voice traffic over a landline when the user is at home or in the office. When in a designated fixed location, the dual-mode DECT/GSM phone would operate in DECT mode with a standard domestic DECT base station, often implemented as a cordless phone base. Separate DECT-only handsets could also be supplied as part of the consumer bundle.

The first commercial DECT/GSM dual-mode service was introduced in the UK in May 1999 under the name of Onephone by BT Cellnet using the Ericsson dual-mode phone. The Onephone did not allow inter-mode handovers (calls that began in GSM mode had to be completed in GSM mode; those started in DECT mode had to be completed in DECT mode), but it allowed automatic roaming when the user was not engaged in a call and moving between different coverage areas (when the user left home, the Onephone automatically switched into GSM mode as the user went out of range of the DECT base station, typically under 300 meters). Operating in the home, the Onephone behaved as an ordinary cordless DECT base–compatible handset, which meant that it could be used to make intercom calls to the other DECT handsets via the base station. The Onephone also allowed manual override, so that if the domestic line was in use, the Onephone could be used to make a call in GSM mode, in effect providing an extra on-demand line.

As a part of the Onephone service, BT also offered the Flexinumber option, when the user received a single number (prefixed 07041 or 07071) that allowed calls to be automatically routed to the user's Onephone, whether the user was at home, at the office, or away from both. A Onephone Flexinumber was also available to consumers who did not use

TABLE 1.3   GSM and DECT Compared

Parameter	GSM	DECT
Range—cell size	Large cells, typically 1–30 km	Small cells, typically 1–500 m
Frequency planning	Fixed, inflexible	Dynamic, adaptive
Traffic per subscriber	~20 mErlang	~200 mErlang
Data capability	Low, e.g., 9.6 Kbps per slot (but higher with GPRS and HSCSD coming)	High, e.g., 28.8 Kbps per slot (but 500 Kbps multislot, and higher with new modulation schemes coming)
Robustness to interference	Moderate	High

BT as their fixed-line telephone provider—in this case they received their normal telephone bill from their usual service provider and, in addition, a Flexinumber bill from BT. A separate bill was also provided by BT Cellnet for the mobile calls made by the user.

The large size of the phones (it was basically two distinct technologies with corresponding chipsets crammed into one handset), the cost of the package, the limited equipment choice (due to limited vendor support, basically having one phone to sell from Ericsson), the lack of a DECT-GSM handoff, and the requirement for two bills (one for the fixed calls, one for the mobile calls—probably due to billing system constraints) caused the eventual demise of the product.

Wireless PBX (MicroTAC/Home Base, Tele-Go [FreedomPlus])

A number of Wireless PBX services were launched using cellular frequencies in the mid-1990s but were ultimately discontinued due to spectrum congestion. GTE launched this service under the name of Tele-Go with Motorola equipment. It was later licensed by Bell Mobility in Canada. Other ISM-band wireless PBX services (often referred to as PBX extensions) have also been around in some form for over a decade.

Essentially this technology required the consumers to install a small picocell-like device in their house. The device appeared as a regular base station to the infrastructure and connected to the MSC via the PSTN line. When a call was placed from home through the home base station, it dialed the network and forwarded the voice traffic to the MSC. At the time the services were launched, the technology underlying this solution was essentially immature and expensive. Also, its reliance on PSTN as a backhaul was far from optimal. Ultimately, the high cost of the home base station, poor services integration, and lack of billing and voice mail convergence contributed to the eventual shutdown of the majority of commercial offerings.

CT2/CT2+, Rabbit, Fido

CT2 was an early cordless telephone standard enabling launch of a commercial service in early 1990 by BT and Hutchison in the UK and Hong Kong under the name Rabbit; in France, under the name BiBop; and in Italy, by Telecom Italia, under the name Fido. The service required the build-out of an infrastructure of public wireless hotspots. CT2 was initially intended to provide a mass-market wireless product with limited mobility at a time when cellular service and equipment were priced too high for casual residential use. The CT2 handsets were less expensive and generally sold as indoor cordless phones that could also be used for mobile telephony over a wide area or just outside of the home (e.g., Fido service was limited to use around the customer's home, to prevent potential competition with mobile services offered by GSM).

Basically users could only make outgoing calls when they were sufficiently close to transmitters located in public sites such as railway stations, the underground, gas stations, etc., just as a cordless phone in the home can operate only up to a certain distance from a base unit. Base stations for this service were even planted on top of bus stops and telephone booths. The handset accordingly required less transmission power than a GSM handset and therefore cheaper batteries than a cellular phone, and it represented a fundamentally different route to mass mobile telephony as an extension of the increasingly familiar cordless phone.

In practice, most of the operators who first took up CT2 licenses decided not to launch a product in the end, as it could be seen as a direct competitor cannibalizing mobile telephony revenues; they opted instead to focus on increasing the number of their cellular subscribers, which were providing better ARPU. Among other reasons for the eventual demise of CT2 was the fact that the CT2 phones lacked interoperability with cellular (either analog or digital), while hotspot coverage was far from ubiquitous.

TDC, Duet, and EnRoute

This group of services was focused around automatic forwarding, often enhanced with simultaneous ringing, and did not require customers to change their fixed or mobile phones. Instead it enabled them to use one phone number for both fixed and mobile services while allowing an inbound call diverted from a mobile line to a fixed line to be charged at fixed-line termination rates when the subscriber was at home. A dedicated server detected if the mobile line had been switched off and forwarded all inbound calls to the fixed line. That is, the customer received an inbound call on the mobile line if switched on, or on the fixed line if the mobile line was switched off. Customers received just one bill and used the same customer support number to resolve issues.

The primary enablers of the services in this group included Intelligent Networking (IN) back-end functionality necessary to blend the fixed and mobile networks through SS7 signaling and a converged billing infrastructure allowing a subscriber to receive a single, "context aware," bill.

This service was launched by Tele Danmark under the name TDC and later by Belgacom as Duet. Duet's initial acceptance was in large part due to simple arbitrage of fixed versus mobile rates. Later in 1997, U.S. operator Ameritech launched a Duet-like service in Indianapolis under the name EnRoute. After some early success, the subscriber numbers ultimately dwindled and the services had to be shut down. Brief descriptions of both Duet and EnRoute follow.

Duet   In its basic form, Duet combined a normal fixed network subscription connected to a digital exchange with an additional subscription to cellular service together with a number of call forwarding and answering facilities. The Duet service allowed the customers to have only one telephone number for both mobile- and fixed-service calls and to get all calls invoiced on the same bill issued by the fixed service. The incoming calls to a mobile number were directed to the mobile handset, when it was on. If the handset was off, the call was directed to the subscriber's fixed telephone. If the fixed phone was not picked up or in an active call, the call was routed to the subscriber's voice mailbox.

EnRoute   More advanced than Duet call forwarding, Ameritech's EnRoute service added a "follow me" feature that simultaneously rang up to three numbers selected by the user. The service also allowed the user—either by phone or through the Web, using the LinkView feature—to place calls, hear or view faxes, listen to voice mail, and change the three phone numbers selected to ring.

Genion, One-Phone DU, FastForward

Genion was launched by UK operator O2 and was the first service marketed to the users around the concept of a home zone with an accompanying price scheme. The service was soon followed by the Korean company SKT's service named One-Phone DU. In this service, the location of the customer was determined by sufficiently closely spaced base stations. Genion's weakness was that the service involved two numbers. Incoming callers had to call the home number first to see if the user was within the home zone. If the user was outside the home zone, callers had to use the standard mobile number.

FastForward   FastForward, introduced by Cingular in October 2003, allowed customers to forward their wireless calls to a landline phone as long as they also subscribed to the SBC landline service. Calls forwarded to the landline would not count against the customers' wireless minutes.

Cingular customers had an unlimited number of incoming wireless calls forwarded to their landline phones. Customers who received a single bill for Cingular wireless and SBC landline services could utilize the service for no additional charge. BellSouth offered the FastForward service when customers signed up for a combined bill and the choice of two other features.

To use the service, the subscribers had to purchase a cradle compatible with their mobile phones. After the initial setup, the service could be invoked by placing a phone in the cradle, which would cause the phone to auto-send a text message to a preset number. The service would work only if there was cellular coverage at the cradle's location. The service was intended only for incoming calls; the wireless phone still had to be used for outgoing calls even while the call forwarding service was active.

FMC Then and Now: Many Services over One Network vs. One Service over Many Networks

Let us now fast-forward to the present time and analyze a fundamental distinction between telecommunications[6] past and present, specifically the change introduced by IP and the global Internet.

ISDN

Despite many changes the industry has undergone during its century-old history, basic governing principles and *physical* properties of transmission media, such as the upper bound of the digital transmission capacity of a link (in bits per second) as a function of

---

[6] The discussion in this chapter mostly applies to fixed communications. However, with the rise of IP as a primary protocol for wireless data and recent strides made by the wireless industry to begin the conversion to mobile VoIP, the conclusions made here are fast becoming applicable to cellular.

the available bandwidth and the signal-to-noise ratio, have never changed (as per the Shannon Theorem [1]). However, the way telecommunication networks and services are delivered and the way operators address market needs and define new competitive offerings have always been a function of *technological* development and marketing strategy.

Coming out of the transition from analog to digital telephony, the ISDN brought the promise of a new transport mechanism allowing for combination of digital voice, audio, video, fax, and data on a single network. The operator would provide customers with network terminations offering either a basic rate interface (BRI [2]) or a primary rate interface (PRI [3]). Services could be offered by using one or more of the digital channels provided by the ISDN network operator. With the ISDN model—thought of as a breakthrough at the time—one network was to provide all services using a single network interface, and the user could access these services only from the premises where this network interface was provisioned.

Also, the ISDN was based on the establishment of TDM (time division multiplexing) circuits between the endpoints involved in service delivery, such as fax machines, video telephones, PSTN telephones, and ISDN routers. It was in essence circuit-switched. This implied quite a degree of inflexibility and inefficiency in bandwidth allocation and, most of all, network resource usage (TDM circuits require dedicated resources), which in turn, manifested in relatively high costs.

Perhaps for these simple reasons the ISDN never became a successful mass-market solution in most countries. It was largely limited to vertical applications such as ISDN backup (an access network that would stay idle most of the time and be turned on only when main service over Frame Relay or ATM would go down). Also teleworkers in some countries using ISDN as a way to forgo analog modems[7] and connect back to the corporate network, at the same time having the ability to call or send a fax.

B-ISDN and ATM

An evolution of the ISDN was Broadband ISDN (B-ISDN), based on the Asynchronous Transfer Mode (ATM) packet-switched technology. ATM again allowed for the delivery of a variety of services over a single network interface (also known as the User-Network Interface, or UNI). The logic behind service delivery based on B-ISDN was still the same as for ISDN, in that virtual ATM circuits were established between customers using services delivered over the B-ISDN network.

However, the Internet Protocol (IP), which, being a ISO-OSI Layer 3 technology, does not care about the underlying Layer 1 and 2 networks, put an end to any hopes of success for B-ISDN. B-ISDN, as a packet technology, probably could have succeeded in terms of more efficient resource usage than ISDN. Nevertheless, its success would imply that all networks in the world eventually converge to use the ATM Layer 2 technology, which clearly was not going to be the case, in large part due to the burgeoning success of the Ethernet in enterprise networking.

---

[7] In Europe this was the primary remote access technology available for teleworkers. In the U.S., however, it was deemed to be too expensive from the beginning and was only offered to VIPs.

In our view, IP became much more successful than many proprietary protocols devised by computer vendors (like the IBM SNA), simply because of its elegant simplicity and the degree of global internetwork interoperability it allows. Ultimately, access to the global Internet to allow for exchange of information, remote collaboration between businesses (via e-mail, file download or upload, etc.), and access to the World Wide Web (WWW), all forced the industry to steer clear of proprietary networking architectures. Consumer network access services were also driven by consumer demand for other Internet Protocol–based applications (such as e-mail and FTP).

Not quite disappearing yet, like ISDN, ATM is also now largely relegated to niche applications such as a Layer 2 technology in ADSL access and cellular backhaul. It never could deliver on the B-ISDN promise and provide technology foundation for global interoperability among the communications & computing (C^2) devices.

IP and the Internet: Breaking Free of Layer 2

This global interoperability has been achieved instead over a network of networks independent of Layer 2 technology. There are no longer any geographical limitations in access to services offered over the Internet.[8] In fact, IP and the Internet represent not only a new way to access services and interoperate networks but a technological (and societal) paradigm shift.

Before the advent of the Internet, a telecommunication service implied a permanent point of access. The PSTN telephone service, for example, could only be accessed through a plain old telephone service (POTS) line offered by a service provider and terminated in a house or an office. Likewise, we could access ISDN services only at the network termination point provisioned by the service provider. Finally the news from news agencies would come only to a fax machine plugged in at a specific location.

Today, the Internet has enabled service portability, meaning that any service can be made available at any access point on the Internet. For instance, we can buy a Voice over IP (VoIP) service subscription from an Internet phone service provider or any third-party provider, get assigned a telephone number, and then carry this telephone number with us while on a trip for business or leisure, and be reachable as long as we are connected to the Internet with the necessary VoIP equipment.

The same applies to our instant messaging service such as Yahoo IM or MSN messenger. We can even access the TV service we subscribe to at our home by encoding the channels and sending the content to a client on our mobile device anywhere in the world where the broadband connection is available. In fact, TV service subscriptions themselves are well on track to be unbundled from the physical access network such as cable infrastructure as IP TV becomes more and more widespread. Potentially any IP multimedia service can be controlled by an IP multimedia service provider regardless of the location where we happen to be and the access technology we use. There is no limit to the number of customer/provider relationships that can be established over the Internet.

---

[8] As long as any Layer 2 connectivity to the Internet is available.

In summary, the fundamental change that the telecom, computing, and media industries have been undergoing during the past few years is the transition from bundles of services accessed over a single network technology to the delivery of a service over any network providing access to the Internet or an IP network.

Essentially, the old model supporting the delivery of many services over one network is being replaced by a new one providing a uniform set of services over a multitude of different networks. Whether a network offers fixed or mobile service, or whether it is wireline or wireless, is immaterial. As such, operators increasingly frequently bundle together different end-user services (telephony, TV, Internet access) and access networks (cellular, WLAN, DSL). New solutions are in the works tying a single telephone number to a user rather than a location, an access network termination point, or a device. More and more we will be able to see the same local TV news at home, in our hotel room on holiday, or even on our mobile device.

The value chain is also shifting in the process. The access network service is still necessary, but it is quickly becoming a commodity. The unbundling of service provision and subscription that the Internet allows makes the access network service a mere price competition play. (Users will pick and choose the cheapest way to get access to their services wherever they are!)

Is FMC Happening?

The shift from the past model built around providing a multitude of independent services over a single homogenous network to a new one enabling delivery of homogenous services over a multitude of heterogeneous access networks is the single most important driver behind the formation of FMC as a field of the telecommunications industry and ultimately a major contributor to its success and wide proliferation in the marketplace. The way the telecom, computing, consumer electronics, and media (content) industries will take advantage of the new model separates winners from losers, failed solutions from successful ones, and popular converged applications from failures.

Indeed, things seem to be falling into place for FMC as of late. For once, migration to access-independent networking—which appeared to be a missing link in eventual FMC success—is happening at an increasingly fast pace. The current service model will of course be dependent on the embedded legacy communications mechanisms until the transition is completed, so a variety of FMC solutions integrating many different types of networking will be required in the meantime.

Second, the market seems to be finally ready. The current services framework has become so overly complex that it must change fundamentally to ensure the industry's long-term survival. Consumers and business users alike demand services to be simple, cost effective, and useful (i.e., not a toy they can use once in a while to show their sophistication).[9]

---

[9] The term casually used in Japan, one of the most technologically advanced countries, to describe gizmo aficionados preoccupied with showing off their technical prowess is "Big Face." Interestingly, showing off one's savvy and technological advancement (in a positive sense) is one of the major driving forces there, where the attraction for gadgets and latest technologies is at levels unknown elsewhere.

It is also essential that users not be required to be aware of technology (is it Wi-Fi? Oh no, is it Bluetooth . . . or maybe wide-area cellular coverage?). They need to receive service enabling a specific desired experience regardless of the type of connectivity instead. The competitive pressure building up within the fixed and mobile industries is also the primary driving factor from a business perspective, as carriers on both sides of the industry strive to invade one another's turf and fend off competition by building integrated and converged voice, data, video, and mobile offerings (also known as "quadruple play" services).

Finally, a strong business case (broadened coverage, increased capacity, decreased cost, and new offerings/pricing plans) will eventually be formulated, leading service providers in both fixed and mobile businesses toward deployment of FMC.

Summary

In this chapter we introduced FMC and defined a foundation for this field of the telecommunications industry, along with the main components of the technology. We also gave a snapshot of the market for the FMC solutions and, what is more important, took the first look at actual user experiences enabled by a few FMC solutions that have been deployed by some service providers in the past. Along the way, we clarified the important differences between fixed, mobile, wireless, and wireline, helping you to better orient in the field. In the next chapter will analyze the FMC business case and its main drivers. In Chapters 3–5 we will further decompose and analyze the FMC into its main components—fixed, mobile, and convergence—and will dive deeper into the underlying technologies.

Chapter

2

Business of FMC

Beware of all enterprises that require new clothes and not rather a new wearer of clothes.

—Henry David Thoreau, *Walden*

No technology has ever been successful without a solid business case behind it, and FMC is no exception. In keeping with the main focus of this book on the exploration of the FMC at large, in this chapter we are going to analyze such business aspects of convergence technology as its ecosystem, drivers and barriers, and addressable market, as well as its segmentation, deployment, and go-to-market (GTM) strategies. The findings of this chapter should form a kind of blueprint, which can help service providers, organizations, and individuals subscribing to services to steer clear of FMC fallacies and pitfalls, to successfully deploy it, and to make good use of it.

New Band in Town

One interesting property of FMC, which sets it apart from many other technologies in the industry, is the diverse and dramatic impact it has—both positive and negative—on various subscriber segments and service provider categories.

For instance, wireless operators may view FMC as a cost-effective (read: cheap) way to improve coverage in certain areas by leveraging "other people's" fixed broadband infrastructure instead of building new cellular towers. For another example, mobile virtual network operators (MVNOs) and VoIP service providers, who often do not own access facilities, may find FMC a sure way to round out their offerings with either mobile or fixed components, allowing them to introduce an array of unique differentiated services to their subscribers.

The dangers (real and perceived) also abound, with FMC often feared by fixed opera-tors as a technology potentially accelerating fixed-to-mobile migration (The threats of which we discussed in Chapter 1). Wireless carriers in turn often see it as a twisty path leading to price erosion, revenue cannibalization, and a potentially dangerous approach to cost cutting, threatening to drive customers and minutes of usage away from cellular infrastructure. (The illustration from *The Economist* in Figure 2.1 nicely sums up the business community's early view of FMC.)

In short, many signs point to the fact that FMC, with all its controversies, is not just a "new band in town" or yet another enterprise requiring new clothes—paraphrasing Thoreau's *Walden*—but rather a technology capable of fundamentally altering the ways in which networks and devices are built and services are offered and consumed.

Despite many roadblocks, perceived dangers, and potential pitfalls, we believe that the fundamental changes FMC introduces to both industry and consumers are mostly positive. As we progress through the chapter, it will become clear why it should be looked at as a new approach to communication, the kind that comes along only a few times in a given technology's life cycle.

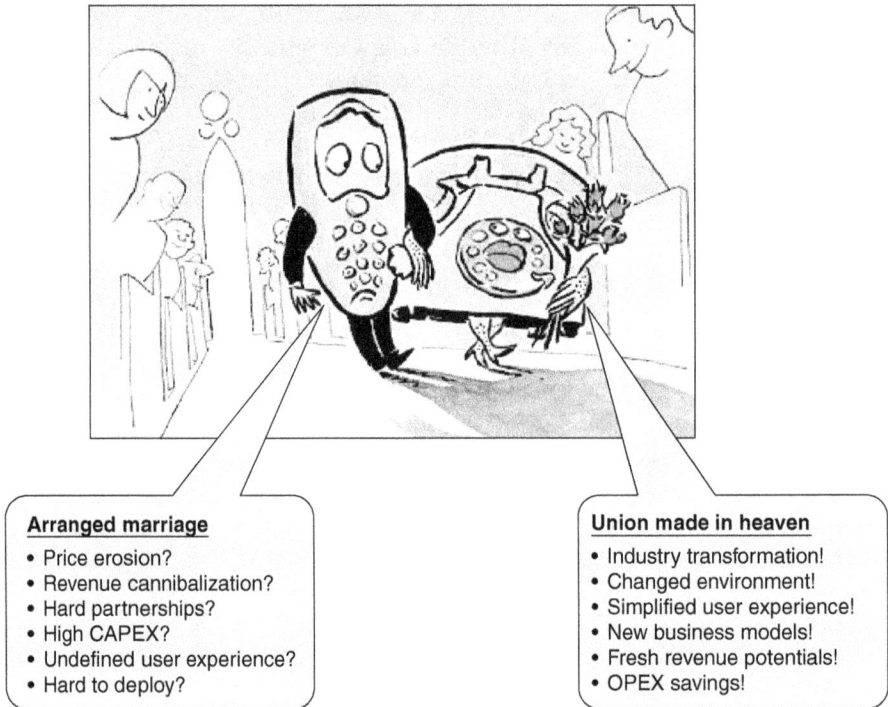

Arranged marriage
- Price erosion?
- Revenue cannibalization?
- Hard partnerships?
- High CAPEX?
- Undefined user experience?
- Hard to deploy?

Union made in heaven
- Industry transformation!
- Changed environment!
- Simplified user experience!
- New business models!
- Fresh revenue potentials!
- OPEX savings!

Figure 2.1   FMC: Perilous matrimony or marriage made in heaven? (illustration courtesy of *The Economist*, September 23, 2004)

FMC Ecosystem

Before sinking our teeth into the FMC business case, it is important to outline the main elements of its ecosystem. Those include service providers, equipment and software vendors, and customers. For the purposes of the discussion to follow, we split the service providers into *mobile* and *fixed* categories, which are further segmented into subcategories as shown in Figure 2.2. Along the same lines, vendors include application, end-user equipment, and infrastructure manufacturers. Consumers or end users can be roughly divided into enterprise, institutional, and residential users. In fact, such a view of the FMC ecosystem is in general applicable to the industry as a whole; however, we are going to use it in a manner best suited to the subject at hand.

Mobile Operators

As in Figure 2.2, mobile operators in our model can be categorized as traditional cellular wireless operators, MVNOs, and Wireless ISPs, or WISPs. The last of these may include Wi-Fi hotspot operators, emerging WiMax providers, and, to a certain extent, players in vertical markets such as microwave, laser, and other line-of-sight last-mile services.

It may be argued that out of all of these groups, MVNOs, not having to deal with the baggage of legacy operational structures, are best positioned to offer FMC services and could put the most pressure on established wireline and wireless carriers to roll out

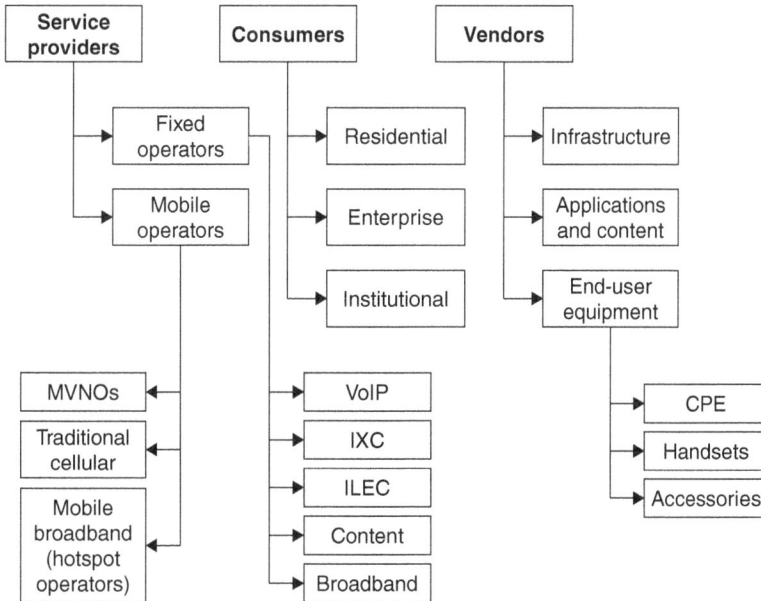

Figure 2.2   FMC ecosystem

similar offers. Indeed, MVNOs are nimbler, less constrained with regulations, and, by definition, more agile than any of the incumbents. Also, being sizably smaller, they are more likely to see FMC as an opportunity to gain market share and offset any coverage disadvantages the networks they rely on might have.

Fixed Operators

This group includes the traditional, heavily regulated and tariffed PSTN players[1] like interexchange carriers (IXCs) and incumbent local exchange carriers (ILECs), as well as less-established VoIP telephony providers employing various operating models to offer the same or more advanced services as their PSTN competitors. Another fixed operator category is broadband networking providers, offering high-speed data service over coaxial cable, DSL, and fiber facilities, which they typically own.

The final group is content providers—until recently almost exclusively composed of cable operators, also known as multiservice operators (MSOs)—delivering entertainment and information content such as TV and audio programming to end users.[2] As in the case of mobile operators, many of the fixed carriers often provide combinations of services to their subscribers. For example, cable operators, with their widely deployed last-mile infrastructure, often take the role of broadband and VoIP service providers. ILECs now customarily provide DSL broadband service and in some cases video content programming through their newly installed fiber to the premises (FTTP) facilities or, if copper quality allows that, via DSL lines. The term *triple play* (for broadband data, content, and voice) is often used to describe the capability to offer such service bundles.

The end game for these multipronged operators is to provide as diverse an array of services as possible to tighten their grip on subscribers. The FMC should offer this group an opportunity to engage in a *quadruple play* as in Figure 2.3, rounding up their service offerings with wireless communications and providing their subscribers a "total" connectivity experience.

Vendors

This group consists mainly of manufacturers of various software and hardware elements to make telecommunications happen, such as:

- Infrastructure components, like core network hardware and software, including switches, routers, and other core functions, as well as radio access network (RAN) components, including base stations and controllers

---

[1] And in some countries, government-owned monopolies.

[2] Faced with the need to make the size and scope of this book manageable, we decided not to cover satellite and terrestrial broadcast communications in this text. Note, however, that satellite service providers play a significant role in both content delivery and downlink broadband services. Also, Internet content providers allowing upload and download of video content should not be forgotten.

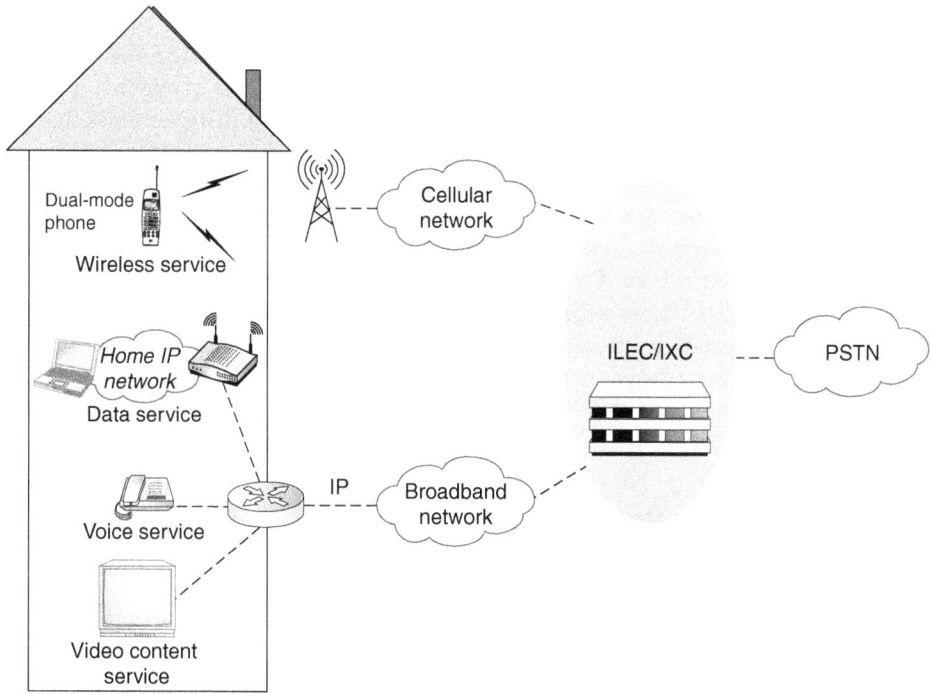

Figure 2.3   Quadruple play

- End-user equipment, including customer-premises equipment (CPE) like cable modems, mobile devices, and fixed phones
- Specialized applications and systems, such as back-end billing and management functions

All of these vendors play an important role in FMC deployment and therefore stand to profit from broad acceptance of this technology. The majority of FMC solutions (such as Wi-Fi/cellular) incorporate all three equipment categories. However, client-based methods might be implemented totally in software and installed on off-the-shelf computers and smart devices.

While it is hard to predict the exact shape that the future predominant modes of FMC deployment will take, it is likely that the bulk of the effort will lie on the shoulders of infrastructure and end-user device manufacturers, with applications vendors playing supporting roles. It also should not be forgotten that the role of system integrators in bringing the solutions together will not be a minor one, as the complexity implied by integrating equipment, applications, and devices from many vendors will be significant. Those able to deliver turnkey solutions will gain a competitive advantage.

Consumers

The main segments in the FMC ecosystem on the "service-receiving end" include residential, enterprise, and institutional (such as government, research, healthcare, and military) subscribers. These markets have different needs but are driven by similar objectives: cost savings on both service and equipment sides, convenience or productivity enhancements, and coverage or reception and potentially service-quality improvements.

Out of the three groups, the residential segment represents by far the most significant target market for FMC in terms of sheer subscriber numbers, while the enterprise and institutional segments present opportunities for higher margins. Although it is hard to predict which of the three subscriber groups will be the most amenable to adopt FMC, the residential segment has recently emerged as the front runner, with the enterprise and institutional environments presenting early challenges to many FMC technologies.

FMC Impact on Service Providers and Consumers

Enabling voice communications over the same fixed networks that handle data transmissions, and converging it with mobile cellular systems to deliver a seamless user experience, is by now recognized as a key strategic advantage by most service providers. Such service offers, however, require putting in place complex infrastructure solutions, creating and managing new handset and CPE programs,[3] and often building new partnerships and alliances for infrastructure sharing, joint development, billing, and roaming. For those fortunate mega-carriers owning both fixed and mobile telecommunication assets, there is the additional challenge to find the way to internally partner on new services without harming the business of the other division. Finally, the need to put in place a workable deployment and GTM strategy goes without saying.

The approaches to solving these challenges—and the nature of the challenges themselves, for that matter—vary widely between service provider categories, market segments, and even geographic areas and the competitive and regulatory landscape in which they operate. While the operators' ultimate long-term goals are essentially similar, both the short-term motivation and the barriers to FMC deployment could not be more diverse.

Fear and Greed

FMC is remarkably full of contradictions, especially for carriers. Take the dual-mode Wi-Fi/cellular, the essential FMC technology, for example. Some cellular carriers are naturally reluctant to deploy this solution because it routes calls normally made over a cellular network through a broadband infrastructure that they often do

---

[3] The CPE complexity is usually associated not with the CPE itself but rather with its management and support. For example, while in most cases standard off-the-shelf Wi-Fi access points can be used for residential FMC, their support strategy is less clear: Would the customer place a call to the wireless carrier, the broadband operator, or the VoIP service provider's technical support in case of service failure?

not control (we will refer to this type of deployment as *arbitrary* or *uncooperative broadband*)—a new, unfamiliar, and less predictable way of handling user traffic.

In fact, many in this camp believe that thus-deployed FMC solutions are harmful for the bottom line, driving the minutes away from cellular networks they had spent millions to build. Other potential pitfalls of FMC (described in detail in the later section "Barriers") include self-competition and revenue cannibalization, price erosion, and risky initial deployments with potentially slow return on investment (RoI).

Others, in contrast, recognize the FMC benefits and associated potential for fat profit margins and are therefore in a hurry to jump the gun and deploy something— anything—out of fear to be left behind and succumbing to competition in case FMC takes off in a big way.

Some fixed operators in turn suspect—not entirely without justification—that FMC has the potential to speed up fixed-to-mobile migration and negatively affect subscriber awareness and perception of fixed communications, essentially turning them into dumb pipe as the user communication experience turns increasingly mobile.

On the positive side, many ILECs, IXCs, and cable operators consider the proliferation of residential Wi-Fi and the widening availability of broadband data service as a chance to counter stagnating subscriber growth and falling voice revenues by moving into new markets with new offerings. They are not discouraged by the growth pains of Voice over Wi-Fi (VoWi-Fi) technology, as the roadblocks are being knocked down, one by one, by vendors with a slew of new devices, infrastructure components, and applications enabling new compelling services and solutions.

In fact, some of these challenges are even being addressed by careful selection of target markets. For instance, many of the early technical limitations of VoWi-Fi—such as security, capacity, QoS, and lack of standards for handover between access points (APs)—may not be as critical for residential Voice over Wi-Fi, making this market a good ground for initial FMC service deployments with minimal risk.

4 Cs of FMC

But what makes FMC so attractive for both service providers and subscribers alike? Primary drivers to deploy FMC can be summed up in what we call the *four Cs of FMC,* as in Figure 2.4. They are, in the order of importance

- **Cost** FMC service spells big cost savings for service providers, which in turn translates into lower prices for both residential and business subscribers. New tariff and pricing flexibility enables new attractive concepts like *home-zone pricing* among others.

- **Coverage** FMC technology can be used to fix gaps in cellular network coverage both in-building and in remote areas without the need to deploy expensive cellular infrastructure. Conversely, it enables fixed operators to offer subscribers wide area service outside limited disjointed areas covered by Wi-Fi access points for complete service continuity unbound by the types of access.

(continues)

- **Capacity** FMC provides a fast and inexpensive way for carriers to increase peak capacity of their networks by offloading traffic to less-expensive access media (from cellular to Wi-Fi, WiMax, or IP-based femtocells, for example).

- **Convenience** FMC dramatically improves service usability and convenience in multiple ways for many user segments, thereby reducing churn and improving ARPU for many categories of service providers.

Secondary drivers, stemming from the primary four, include new forms of competition, the ability to introduce new applications and services, which otherwise would not be possible, and access to new, previously underserved or "un-servable" market segments. Last but not least, FMC affords service enhancement and expansion potentials, both strategic and tactical, for example, enabling the coveted *quadruple play* for fixed operators or the ability to enter the MVNO business and extend their offerings to the wireless medium for VoIP service providers.

FMC Drivers and Barriers

This section offers more detailed analysis of both positive and negative impacts of FMC on service provider business along with an in-depth discussion of its drivers and barriers around the four *C*s.

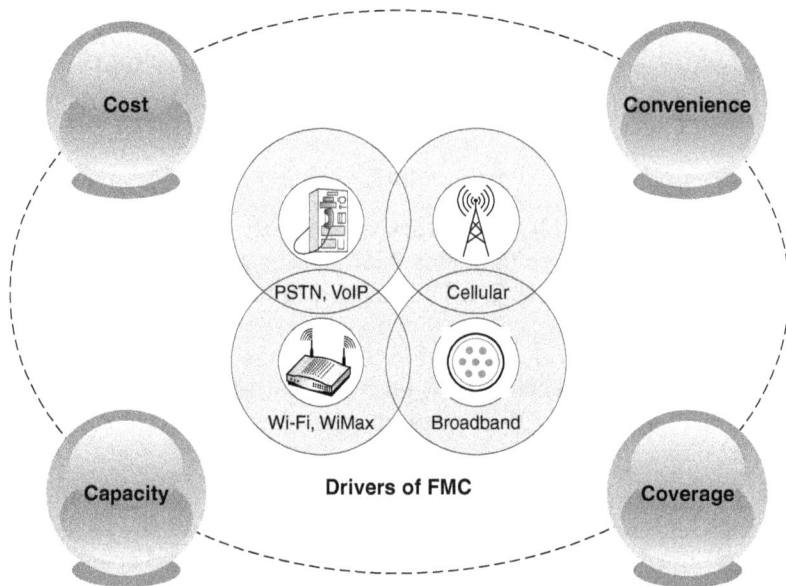

Figure 2.4   The four *C*s of FMC

Table 2.1 summarizes the impact of FMC on service provider business in a framework of a *Strengths, Weaknesses, Opportunities, and Threats (SWOT)* model. From a look at this table, it becomes apparent that the FMC impact on service providers varies widely within category.

Using this table to facilitate the discussion, let's now analyze this intricate tangle of opportunities and challenges.

TABLE 2.1   FMC Impact on Service Providers

Impact	Mobile Operator			Fixed Carrier	
	MVNO	Traditional Cellular	Mobile Broadband	VoIP	PSTN
Strengths					
Cost	X	X			
Coverage	X	X	X	X	X
Capacity	X	X			
Convenience	X	X	X	X	X
Opportunities					
Quadruple play	X	X	X	X	X
Access to new subscriber segments	X	X	X	X	X
New services and applications	X	X	X	X	X
New business models and markets	X	X	X	X	X
Churn reduction	X	X	X	X	X
Weaknesses					
Components complexity	X	X	X	X	X
Difficult partnering		X	X		X
Cannibalization		X		X	X
Technology immaturity	X	X	X	X	X
Threats					
Price erosion		X			X
Diminishing needs	X	X	X	X	X
Subscriber inertia	X	X	X	X	X
Weak initial business case		X		X	X
Regulation	X	X	X	X	X
Operator inertia	X	X	X	X	X

Drivers

This section analyzes both FMC strengths and the opportunities it offers to both fixed and mobile operators and subscribers.

Cost   Cost reduction may be the most important benefit of FMC for both service providers and subscribers alike. Indeed, FMC can be used by wireless carriers to substitute more expensive types of access and provide coverage at significantly lower levels of both operational and capital expenditures (OPEX and CAPEX). Thus a network of Wi-Fi hotspots may carry calls and data sessions for much less than a cellular network (cheaper backhaul, unlicensed spectrum, etc.). Many of the operators' findings also indicate cost savings to be of primary interest to both their residential and business customers. This equally applies to fixed operators for whom FMC represents the best route to offering quadruple-play service bundles, which makes it possible to lower the total bundle price to consumers.

The FMC cost savings can be passed on to the consumer through an approach increasingly referred to as *home-zone pricing*. This concept allows operators to offer subscribers variable rates based on their location and/or type of network access. Main enablers of this scheme include the ability to accurately detect subscriber location and mode of connectivity and appropriately update the back-end systems capable of differentiated handling, provisioning, and billing based on these and other parameters determined by the operator or (in some cases) by customer preferences.

For instance, an operator with an FMC solution in place may offer a service plan with standard per-minute cellular service charges in all locations except for a home zone. In the home zone (where the specific Wi-Fi access point or femtocell is installed), the subscriber will be able to use unlimited minutes for a flat fee or for free. Alternatively, a service provider may offer two types of per-minute voice and data rates with the higher rate in effect in the area covered by cellular service, and the lower rate applied when the subscriber is connecting through one of the participating Wi-Fi hotspots.

On the enterprise side, FMC allows operators to consolidate multiple existing communications services such as cellular and fixed PBX telephony, paging, messaging, and in some instances even two-way radios (for example, by replacing the fleet of walkie-talkies with modern push-to-talk cellular phones), and therefore save costs not only on service but also on CPE and end-user devices. Here convergence can also provide a cure for some often wasteful (but necessary) practices often employed by workers trying to improve their reachability and simplify their communications.

Many office workers, for instance, often permanently forward their fixed lines to their mobile numbers to be reachable by customers and colleagues calling their fixed numbers while they are not in the office (even authors of this book occasionally find themselves among those guilty of such malfeasance). With an FMC solution in place, this would not be necessary, as members of the organization would use a single number for both fixed and mobile communications with the back-end infrastructure selecting the most appropriate access, thereby providing significant cost savings to organizations and simplifying communications with customers and partners.

The equipment-related cost saving is also an important driver for the consumers, who would need to buy fewer different devices, instead relying on ubiquitous multi-mode handsets doubling up as cellular and cordless in the home or office. With FMC service, consumers would need to purchase—and also charge, maintain, and learn how to use—fewer devices, or, at least, fewer different *types* of devices, or, at the very least, fewer devices with *dissimilar functionality*.

Coverage   Out of all other convergence "Cs," this one is directly related to the subscriber's ability to actually *receive* service. Indeed, the first question asked by customers and organizations' IT departments signing up for telecommunications service contracts is "Is there coverage in my primary residence or place of business?" The consumers' and businesses' demand for better coverage is a direct response to often inadequate quality of cellular service indoors or in remote areas of the countryside characterized by low user density, and as such low investment in outdoor infrastructure. The magnitude of this issue varies between geographies and networks, but the problem is generally present in some form in most markets in both metropolitan and rural settings on all continents.

It is easy to see why for many wireless operators—especially those outside Western Europe and East Asia[4]—coverage improvement became one of the most important motivating factors behind FMC deployment. With this technology, consistent service in vast remote areas with low population can be made available literally overnight. Furthermore, FMC can be the only way to provide mobile service in-building where cellular signals may not reach through the building structure under any circumstances.

Instead of building out expensive terrestrial infrastructure of cell sites and dedicated backhaul network in areas with low population and consequently slower rates of RoI and high OPEX, operators can easily satisfy connectivity demands by offering service that provides a similar user experience through their existing broadband network access. Statistically, the majority of the mobile calls are made indoors and often from only one or two select locations for a given subscriber account, such as the home and office, as illustrated in Figure 2.5.

Likewise, for landline operators, FMC is a perfect vehicle to un-tether their service from a fixed location, making it *location independent* and therefore extending it to cover wide areas. This can be achieved by striking MVNO-style contracts with wireless carriers, tying their VoIP or in some cases even traditional PSTN infrastructure with the cellular network. The resulting service and subscriber equipment such as mobile phones would still carry the familiar branding of a fixed operator, such as BT Fusion, Vonage Mobile, or Skype Mobile, thereby minimizing the potential collateral effect of escalating fixed-to-mobile migration.

---

[4] In those regions and other highly populated geographies such as metro areas, other FMC drivers such as pricing and user experience improvements take precedence.

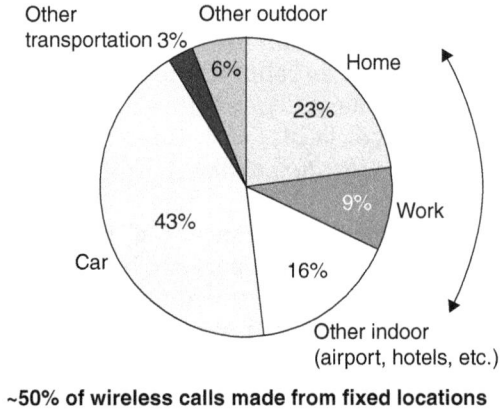

~50% of wireless calls made from fixed locations

Figure 2.5   Mobile calling patterns (source: ABI, Visiongain, Instat)

Capacity and Dynamic Bandwidth Management   Capacity improvements and bandwidth management capabilities offered by FMC are especially valuable for wireless operators using licensed spectrum, a limited resource. In fact, we believe that deployment of FMC solutions combining cellular technologies operating in public licensed spectrum, coupled with a short-range wireless access technology such as Wi-Fi or Bluetooth, or a wider-area wireless technology such as WiMax, is practically unavoidable. The ever-increasing wireless traffic volumes driven by mobile data communications and the rise of services such as media streaming, content sharing, and broadcast will eventually surpass the licensed spectrum capacity even as the air interface utilization efficiency improves with the introduction of next generation of cellular systems (see Figure 1.3 in the preceding chapter).

Even today, by operators' own estimates, best-effort[5] video streaming over High-Speed Packet Access (HSPA) or 1x Evolution-Data Optimized revision A (1xEVDOrevA or DOrA) high-speed cellular networks by only five or six users can consume practically all the capacity of a single cell.

The ability to use alternative access networks provided by FMC offers wireless carriers a simple and elegant way to allocate traffic between a limited-capacity cellular network and higher-bandwidth alternatives (such as a combination of fixed broadband and Wi-Fi or WiMAX), depending on time of day, network load, and other parameters. For example, a wireless operator can provision its customers' multimode devices to default to Wi-Fi mode (whenever available) during nights and weekends[6] or certain peak times, thereby

---

[5] No guaranteed QoS.

[6] Although it seems counter-intuitive, offload at off-peak times makes sense in some areas where the off-peak network capacity is not sufficient to cater to off-peak-time traffic increases due to "all-you-can-eat nights and weekends" pricing plans.

reducing the load in the more expensive cellular network and therefore the need to build the capacity necessary to accommodate peak-time use spikes. Network offload or dynamic switching based on network utilization is also possible, providing a general capability for operators to continuously monitor and conditionally *throttle* overall network traffic.

It must be noted that the importance of network offload varies widely among wireless carriers within the same category and even the same country. That is, the operators with highly developed mobile infrastructures are more capable of satisfying demand and usually more interested in maximizing the short-term return on their investment by fully utilizing the capacity of their networks. Those operators often tend to view alternative access mechanisms, such as Wi-Fi, as a threat and a way to price erosion rather than an OPEX improvement opportunity. Conversely, owners of less highly developed networks with oftentimes spotty coverage or limited capacity, especially in metro areas, are more likely to see potentials of FMC to essentially become a low-cost, on-demand extension of their networks.

Convenience   This driver is a direct consequence of the modern telecommunications usability issues and user frustration at dealing with multiple, often vastly different devices, such as fixed and mobile phones, and services, like fixed and mobile voice plans, multiple separate voice mails, disparate contact lists, incompatible supplementary service, and others (see Figure 2.6). The resulting subscriber dissatisfaction is equally damaging to both fixed and mobile operators.

The carriers that are capable of figuring out the winning formula (user satisfaction = unified user experience + homogeneous service + multimode/multifunction common UI devices) and converging disparate services stand not only to stimulate use and improve average revenue per user (ARPU) but also to reap all the other associated benefits such as reduced churn and improved rates of new customer acquisition.

Both device and service integration play an important role in this process. Address books can be synchronized among all devices in the suite, and so can many features such as the voice mail (VM) message waiting indication (MWI). Whether it is a single multimode device capable of communicating via multiple networks (which, for instance, can be provisioned to latch on to the one with better coverage) or a family of different devices governed by the same rules and supporting similar UI and applications, the user experience will be uniform, contributing to reduced complexity. With these streamlined services, the individuals and corporations alike no longer have to deal with multiple billing statements, different provisioning procedures, and most important, different support departments. Thus converged solutions will provide feature parity between fixed and mobile service components.

Access to New Subscriber Segments   Quite simply, FMC lets service providers target new user segments, which they could not previously serve. For example, when customers decide to sign up for a new service contract with a particular operator and find out that either their home, office, or other locations of interest are not covered by the existing mobile service, they typically walk away. With the FMC system in place, they will instead be promptly offered an alternative (for example, femtocell or Wi-Fi/cellular)

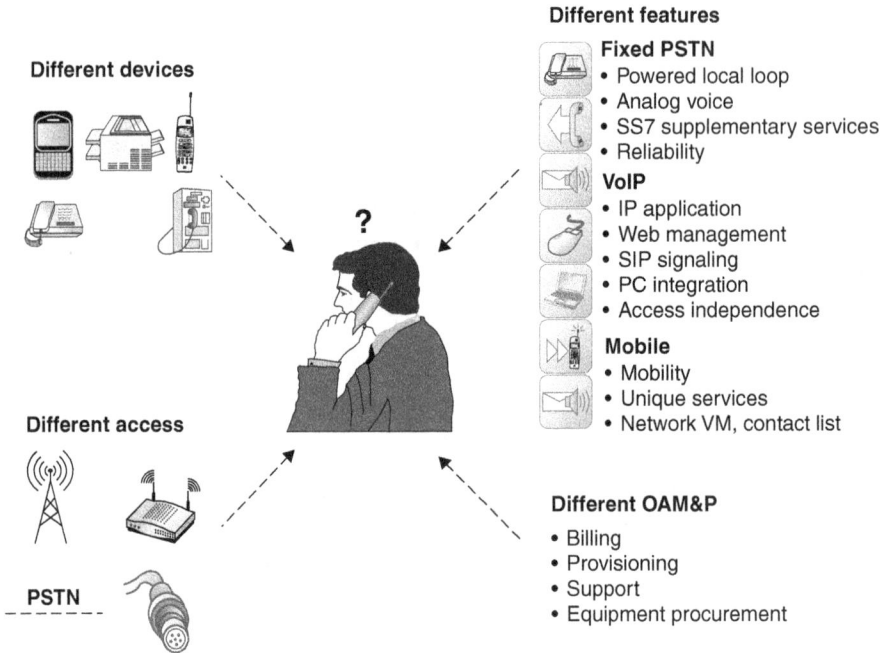

Different devices

Different features

Fixed PSTN
- Powered local loop
- Analog voice
- SS7 supplementary services
- Reliability

VoIP
- IP application
- Web management
- SIP signaling
- PC integration
- Access independence

Mobile
- Mobility
- Unique services
- Network VM, contact list

Different access

Different OAM&P
- Billing
- Provisioning
- Support
- Equipment procurement

PSTN

Figure 2.6 Subscribers and their discontent

solution utilizing their existing broadband access. Alternatively, when landline operators or MSOs offer quadruple play, they are likely to attract new subscribers interested in their integration offerings.

New Services and Applications   Potentially, the ability to route telecom traffic through alternative access channels with different characteristics (such as a secure, mobile, and highly available cellular network or a higher-speed, inexpensive combination of broadband and Wi-Fi or WiMax) enables operators to offer new, media-specific types of applications, which would not be possible without FMC.

FMC also offers ways to drastically enhance existing applications and services to utilize the new capabilities of the resulting "super network." For example, *fixed* operators can offer "follow me" broadcast and media sharing, extending fixed services into the wide area (including coveted "three screens" solutions allowing streaming content to TV, PC and mobile device screens depending on coverage, user location and preferences, and other conditions), mobile operators in turn can offer content download applications, which can be provisioned to synchronize with network servers only when in Wi-Fi, thereby preventing cellular network overload.

Competition   Can we consider competition, even forced, a legitimate driver of a new technology? Taking into account the current state of telecommunications, we answer "yes"!

In the context of this book by competition, we mean something much broader than good old competition with fellow operators in the same market segment or niche. Nowadays, service providers of all denominations must be prepared to deal with much more serious threats from virtually all other carriers getting on the FMC bandwagon:

- Traditional mobile cellular operators
- Broadband service providers
- Cable operators (MSOs)
- Traditional PSTN IXCs and ILECs
- Wi-Fi hotspot operators
- VoIP carriers
- Nascent WiMax service providers

Even local grocery stores, public libraries, and municipalities can now become competitors, offering telecommunication services that in many cases are *free*. It is hard to compete with free!

In fact, as the convergence arms race unwinds, as IP becomes the de facto network protocol for data, voice, and media, and as technologies like IMS go mainstream, it becomes increasingly easy to treat the infrastructure and access networks as part of the same dumb pipe, quickly diminishing the access provider's (who often tend to create "walled garden" service and treat their customers as captive audience) role in the services delivery ecosystem.

Another form of competition faced by operators is the one created by their own subscribers! To address it, "big telecom"[7] may soon have *no option* but to offer FMC in some market segments. The consumers—having tasted the forbidden fruit of access-independent VoIP service coupled with widely offered (and quickly getting down in price) high-speed mobile data and highly capable "smart" devices—will expect and even demand converged service from their providers.

This phenomenon, in fact, extends the concept of *access independence,* described at the end of the first chapter, to *service independence*! If converged services are not offered, consumers, especially those technically inclined, will be engaging in grassroots "guerilla" efforts to create their own homegrown FMC solutions, having all the building blocks at their disposal, including newly emerging intelligent devices with various voice clients and automatic network selection capabilities, as in Figure 2.7.

One could go even further and foresee the rise of the integration industry addressing the convergence needs of less technically savvy users—representing the vast majority of the mobile subscriber base—helping them to satisfy their craving for simplification and cost cutting.

---

[7] Alluding to "Big Pharma," a term often used to describe the pharmaceutical empire, often caught exploiting their consumers.

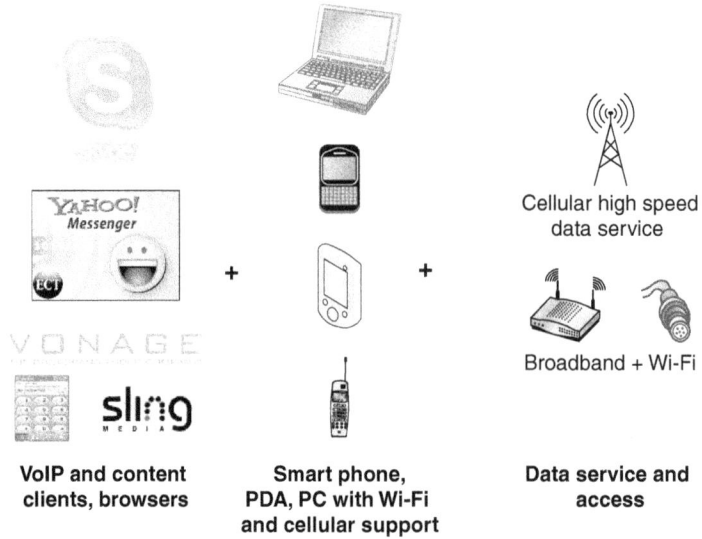

Figure 2.7   Homemade convergence

Barriers

As is often the case with many complex technologies, the barriers to FMC adoption are closely tied to its drivers. While it is not set in stone that introduction of FMC service will have an immediate positive impact on a service provider's business—especially with numerous barriers with potentially negative impact discussed in the text that follows—it is only a matter of time before bottom-line improvements will become noticeable. Indeed, the barriers may play a role in slowing down or limiting the FMC RoI for certain service provider categories or market segments, but in our view, none of them looks like a showstopper.

For the purpose of the discussion, the barriers to FMC adoption in general can be divided into two types: those to do with the technology and those related to business issues.

Technical Barriers   Technology is often blamed on slow pace of FMC introduction. It has, indeed, played a limiting role in many of the past convergence attempts,[8] with the infra-structure, end-user equipment, and lack or immaturity of standards equally contributing to failures. Even at the present time, years past those failed initial attempts, the components of a typical FMC solution remain difficult to implement. While the technical barriers vary between different categories of service providers, the main hurdles include:

- *Device disparity* (creating an integrated strategy incorporating multiple types of devices)

---

[8] See Chapter 1.

- *Transparent mobility* (including both micro- and macro-mobility or inter-access roaming and in-call handover)
- *Functionality parity* (or ability to replicate existing services, features, functions, and most important, characteristics) among different fixed and mobile access mechanisms

The majority of FMC technical challenges has to do with the historically different nature of fixed and mobile communications and associated differences between their services, features, and functionality framework. This disparity is also due in part to the differences between fixed and mobile devices, vastly different infrastructures, and heterogenous fixed and mobile markets, which evolved in unique directions over the past few decades.

End-User Devices Disparity   Let's first take a look at the barriers caused by the end-user equipment. While the handset form factor nowadays might not be much of a contributor to a problem, handset market segmentation (read: pricing), functionality, and feature content certainly are. In the early years of telecommunications, ILECs used to offer their subscribers bundles including both telephones and service. That model changed as independent phone manufacturers entered the market and forced service providers such as AT&T in North America or BT in the UK to either completely or partially exit the end-user equipment business.

This also resulted in commoditization and consequent severe price erosion of wired and soon cordless phones, which are often sold today for less than a monthly service charge for a typical user (and that's often for a bundle of up to three cordless handsets and a base with an answering machine). There are, of course, exceptions to this trend. Much higher-tier products, including high-end residential and small office–home office (SoHo) systems and those targeting businesses and tied to office private branch exchanges (PBXs), offer a broader array of features and also command higher price premiums.

VoIP phones in particular are starting to incorporate features that until recently were found only in high-end mobile phones, such as advanced color graphics display, wallpapers, customizable menus, and extended contact lists. Those examples, however, are just that, the exceptions, with the mainstream market still being driven by price rather than features.

Fast-forward to mobile telephony. The early pricing advantage of fixed phones was more than compensated for by mobile devices' greater feature richness and utility. As a result of commoditization and despite having a head start of approximately 60 years, fixed phones remain hopelessly behind their mobile counterparts in features, capabilities, quality of construction and packaging, and overall sophistication.

In fact, the latest mobile phones and especially *smart phones*—advanced devices usually running open-platform operating systems—have successfully blurred the line between different functions. A modern smart phone now embodies a communication device; a personal digital assistant (PDA); and a minicomputer sporting multiple interfaces, complex operating systems, and development environments (such as Windows Mobile, Symbian, Java, BREW); along with multimedia capabilities rivaling those of full-fledged desktop or laptop computers of just a few yeas ago. Many now even support multiple color screens, cameras, antennas, and wireless communications options.

The gap between the two classes of devices is evident; how a given FMC solution can bridge it is less so. In general, a multimode FMC device that can perform equally well as a mobile device in a wide area and as a fixed device in a home or office environment would face many challenges, both ergonomic and connectivity, as in Figure 2.8. Volume producing mass-market converged devices is not an easy task; however, unlike usability, most of the technical challenges in such devices either have been solved or are in the process of being addressed in next generation, as device vendors are aggressively working with chipset manufacturers and software companies.

Producing a converged multimode uber-handset, however, solves only a part of the problem. A more difficult task facing service providers and vendors alike is building a truly user-focused converged *device strategy* (providing high levels of usability and

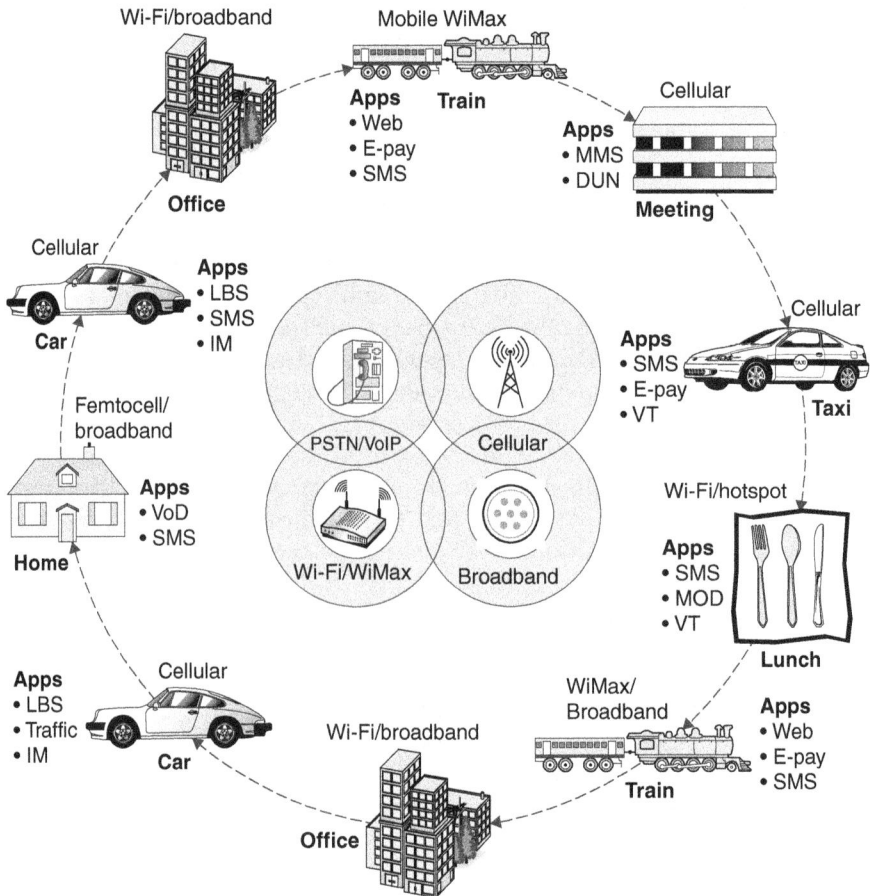

Figure 2.8   A day in the life of an FMC device

enabling senselessly connected experience) incorporating all *three* types of handsets currently in production:

- Conventional mobile phones and data devices
- VoIP and PSTN fixed phones and portable computers
- Converged multimode mobile devices

The importance of such solutions is hard to overstate. Historically, the solutions allowing subscribers to avoid replacing their existing equipment were most successful. Take, for example, the triumphant worldwide advance of VoIP services. VoIP only began booming when the major industry players made a decision to allow subscribers to use their existing PSTN phones connected to routers via terminal adapters instead of requiring them to switch to new VoIP phones. Letting users keep their old trusty devices and thus preserve their familiar experience, made it easy for them to switch to the new type of telephony.

A solution like this sounds contradictory. An astute reader may ask: "On one hand you are talking about new FMC devices; on the other you are touting allowing users to keep their existing equipment. How is it possible to integrate old analog phones with a complex state-of-the-art FMC solution?" That is a good question. There are ways ... but wait till Chapter 6, which focuses on user experience, to find out the answer.

Transparent Mobility   Cellular wireless networks were designed to provide voice service to mobile users, with seamless in-call handover and idle roaming being essential functionality of even early service offers. Wireless LAN technology was introduced to replace wired media for local area data communications, with not only handover and roaming but even basic voice not being requirements until very recently. That example explains the general difficulty of building an FMC solution supporting transparent mobility including both handover and roaming between different types of access networks and traffic types. (Table 2.2 provides the example of the inherent differences between requirements of supporting voice and data over wireless IP facilities.)

It is no wonder, then, that both the vendor community and standards bodies have mounted significant efforts to address voice-related mobility requirements in nonmobile networks.

TABLE 2.2   Voice over IP vs. Data

Parameter	Voice	Data
Delay tolerance	Minimal	High
Mobility	Frequent	Limited
Security concerns	Moderate	High
Client complexity	High	Low
Data pattern	Constant	Bursty
Acceptable error rate	Moderate	Low

Multiple solutions were developed using different approaches for handover and roaming support. Examples include PBX-based forwarding schemas for enterprise, femtocells, and Wi-Fi-based solutions. The latter include UMA/GAN, supporting GSM radio access emulation and tunneling and core network–based approaches such as the IP Multimedia Subsystem (IMS)—more on those in Chapter 5.

Voice over Wi-Fi solutions for one must be able to support in-call handover between access points in a way at least comparable to cellular hard or soft handoffs to make the VoWi-Fi telephony user experience equal to that of cellular telephony. Telephone users are simply more mobile than data users, as it's easier to walk and talk than it is to walk and type. Therefore, especially in enterprise environments, they are likely to roam between APs more frequently, requiring seamless, low-latency handoff technology.

For example, the Fast Roaming Study Group within the IEEE 802.11 Working Group is currently considering the imposition of a maximum amount of time allowed for a handoff, and may set the maximum at 50 milliseconds. As of today, the ability to support standards-based fast handover of calls between APs is still an issue, while proprietary solutions abound.

Functional Parity   Let's now turn our attention to challenges of achieving parity in features and functionality between fixed and mobile components of FMC. As mentioned before, fixed and mobile networks were created to enable different user experience, and have evolved to serve different market needs. The mobile service offerings traditionally supported features needed by subscribers on the move and were therefore fine-tuned to common (small) mobile device form factors. For instance, voice mail in cellular is almost always network based, as are device time synchronization, conference calling, and other supplemental services. Features unique to cellular include text and media messaging, integrated data access, location-based services (LBS), and recently various media access and sharing solutions.

In contrast to mobile, the fixed telecom services were traditionally geared toward the needs of fixed subscribers and maintained strict delineation of voice, data, and content delivery (even today, when all types of traffic are delivered over the same physical media as a part of the same contract), in most cases[9] requiring separate end-user devices and equipment. Basic fixed voice service is typically limited to just that: voice communications. However, this situation is changing to a degree, in recent years, with the introduction of advanced services for both PSTN and VoIP, often including features matching those of cellular service, such as network-based PBX (called *Centrex*), call waiting, and call forwarding.

So the challenge faced by the FMC service providers is to provide users accustomed to different properties of fixed and mobile service environments with a seamless experience incorporating the most attractive features of both. Ultimately the converged solution must also enable a comparable user experience when it comes to service quality,

---

[9] In addition to dedicated handsets, IP-based voice services typically can be accessed by using a computer with the appropriate headset so that the convergence is happening in this area of fixed communications.

including latency, security, speech quality, and other parameters. Many of these challenges lie with the technology itself and more often than not are predicated on specific deployment environments.

Latency in VoIP is directly tied to acceptable audio quality resulting from minimized network delay in a mixed voice and data environment. Generally, voice communications require low latency, acceptable transmission error rates, and minimal signal loss. VoIP communication, being a real-time application, can tolerate relatively low delay thresholds of approximately 200 milliseconds (ms). Anything above that may become noticeable to end users.

The way to guarantee the necessary voice traffic prioritization and acceptable latency figures in a data network is to ensure the delivery of well-defined and predictable levels of QoS. Ethernet, wired or wireless, was not designed for real-time streaming media or guaranteed packet delivery. Congestion on the wireless or wireline network, without traffic differentiation, can quickly make any kind of voice transmission unintelligible.

For example, making Voice over Wi-Fi a part of an FMC solution requires support of a security framework at least as robust as the one used in cellular communications. This fact raises a number of concerns, since many of the standard Wi-Fi security mechanisms do not provide an adequate level of protection. As with data over WLAN, Wi-Fi telephony can leverage existing Wi-Fi security methods such as WEP (Wired Equivalent Privacy), WPA (Wi-Fi Protected Access), and VPNs. We will explore these and other related subjects in depth later in Chapter 5.

Business Barriers   It may be argued that the FMC business barriers are more challenging to overcome than the technical ones. A decades-old competition and the ensuing antinomy between fixed and mobile service providers as well as inherent subscriber inertia are partially to blame. This competition is deeply rooted and is evident in frequent tensions among landline and mobile divisions of integrated telcos and slowly progressing partnership agreements between independent fixed and mobile service providers.

Mobile operators are not very well prepared to conduct business in the fixed world. This is equally true for Cable MSOs, ILECs, and VoIP service providers attempting to expand their offerings into unwired wide area markets. For example, when VoWi-Fi or femtocell technology is used, wireless carriers driving these initiatives must answer questions of fixed facilities ownership, such as

- How would we distribute and support CPE required for FMC solutions?
- Should we rely on partnerships, with broadband providers or offer our service over non-cooperative broadband connections?[10]
- Should we build and own the VoIP infrastructure, or should we rely on a partner and only provide convergence infrastructure and back-end support for billing and provisioning?

---

[10] We like to term this type of offer as BYOB, which stands for Bring Your Own Broadband—the approach widely used by new-age VoIP service providers such as Vonage and Skype.

Fixed carriers and MSOs also must decide how to structure partnerships with wireless carriers and resolve the subscriber ownership, branding, and go-to-market strategy questions. They also should be concerned about escalating fixed-to-mobile substitution, as their new FMC offerings might make it even easier for consumers to switch to use mostly wireless networking and only use fixed occasionally.

These questions are difficult to answer, and it will take some time for the players in both fields to define the new service delivery structures, as they work closely together to realize the true FMC potential to add to their bottom line. FMC, like no other technology, has the power to force *fixed* and *mobile* service providers to come together (in part under the threat to become obsolete and eventually succumb to substitution, become dumb pipes, or engage in damaging price wars). Those willing to change and able to successfully do so, however, will have to make hard decisions about many aspects of their relationships, such as

- Revenue recognition and sharing strategy flexible enough to cover all usage patterns and deliver an array of features fixed and mobile consumers have come to expect

- Service and support roles and responsibilities (My call dropped when I was in a Wi-Fi hotspot. Where do I call for support?)

- Pricing models including home-zone pricing and striking a compromise between wireless and wireline charging and billing practices (unlimited calling vs. per-minute charges, messaging, supplementary services, etc.)

- Subscriber ownership (Does a new subscriber belong to the wireline or wireless arm of the partnership? Who assigns numbers? Who provides the first line of customer service and support?)

- Labor relations (the unionized fixed telecom workforce vs. the typically less-regulated wireless industry vs. VoIP carriers who might not even be located in the country where service is delivered)

Competition   Competition can also be a powerful barrier to FMC. What? Competition again?—you might ask. Didn't we just establish that competition is a good thing—a driver in fact? That's true, but competition might be both a benefit and a kludge. Turning the argument in the drivers section around, FMC is the surest way for operators from different playing fields to intrude on each other's turf. Therefore, service providers in many cases are reluctant to strike agreements that potentially enable their recent competitors (turned partners) to steal their subscribers.

Even service providers with both fixed and mobile assets, most often maintaining separate business units, are not immune to this fear, and for good reasons. The potential dangers of the shift from fixed to mobile are especially visible when one analyzes subscriber growth numbers for such converged carriers. Introduction of an FMC solution can negatively affect the already shaky bottom line of a fixed business unit (and with it the executive bonuses).

One possible way to address this roadblock may be to explore the alternative deployment routes and rely on partners from the "outside," competing for different segments, such as hotspot operators, or limiting offers to non-cooperative broadband deployments. Another solution would be internal realignment and convergence of the organizational structures and changing business units' or divisions' objectives toward improving on the introduction or support of convergence.

End-User Inertia   The end-user inertia (or maybe lack of education) is another difficult hurdle to overcome. The goal of a good FMC solution is to deliver all the benefits to the end user while being essentially transparent. While using most FMC solutions requires minimal or no training, the process of switching in many cases poses a challenge even if the technical side of the switch had been thoroughly ironed out by service provider. That is, the majority of FMC solutions such as Wi-Fi/cellular or femtocells require an effort from the user to initiate the service and install the equipment, which is above and beyond that needed to deal with signing up for separate fixed or mobile services.

For instance, the subscribers of both fixed VoIP and mobile service wishing to switch to a dual-mode FMC may need to perform additional steps to provision their new dual-mode handsets with both cellular and VoWi-Fi networks and install new CPE such as specialized routers or, in case of an enterprise, new PBX equipment. The service may also require the user to switch from a current provider on the VoIP or cellular side.

One obvious way to address this potentially troublesome issue is to further simplify and document the switching experience, making it as easy as possible and, ideally, similar to switching to a non-FMC service. As such, operations comparable to those used for device management or over-the-air handset programming in CDMA or SIM updates in GSM can be employed to invoke Wi-Fi bonding and VoIP-side provisioning. The roles of documentation, advertising, and infomercials in ensuring an easy switch and a positive user experience therefore become especially important.

Deployment and Go-to-Market Strategy

This section takes a brief look at possible FMC deployment scenarios and go-to-market (GTM) strategy alternatives for FMC deployment with the objectives of minimizing initial risk and improving both short-term RoI and long-term potentials for service providers.

Mobile Operators

Mobile operators in developed markets are increasingly facing the subscriber growth slowdown and looking for alternative strategies to improve their bottom line, such as increasing average revenue per user (ARPU) through introduction of novel data services. The initial FMC rollout campaigns in such regions should be strategically focused on addressing only one specific subscriber segment with a few well-defined top needs.

The examples include single urban professionals wishing to reduce their monthly payments and ready to switch to a single device, exurban households looking for at-home coverage improvements, or businesses looking for more reliable mobile coverage on the factory floor. With time, the focus of the rollout can be expanded to broader markets by focusing on secondary subscriber drivers and alternative segments.

In the areas with good coverage and no capacity issues, the promotional programs directed at reduction of long-distance and international calling costs may be used to encourage early FMC adoption. Once the users get a taste of what FMC can offer in addition to reducing their monthly bill, other benefits such as convenience and improved indoor service quality will become more apparent, helping to drive the momentum aided by the word of mouth.

Phased targeted deployment strategies can be equally well applied in other regions such as developing markets or remote locations, by focusing on appropriate sets of drivers. Potential FMC marketing campaigns there can concentrate on eliminating coverage gaps and providing multifunction equipment that can double as mobile or cordless phones, depending on the area in which it is used.

Fixed Operators

Fixed operators—both VoIP providers and ILECs—can also benefit from carefully targeted, phased FMC deployments. For VoIP providers, mobilizing their existing fixed service is a logical next step in the quest to take market share from incumbents. One possible way for VoIP operators to achieve this objective is to strike an MVNO-style partnership with an open-minded second- or third-tier wireless carrier.

Ideally, under the terms of such a partnership, the VoIP operator would also retain the subscriber ownership (the prerogative to assign subscriber numbers from the available pool, provision billing, and to provide the first line of support). Once relationships are established, the VoIP operator would resell and market the wireless service under its own name and be responsible for wireless equipment distribution, branding, and customer care.

Under the MVNO service, the user equipment would be distributed and provisioned in a way best suited to support a few specific business models. For example, for Wi-Fi/cellular FMC service, the VoIP operator–branded multimode handset would be provisioned to fall back to cellular only when Wi-Fi access (whether through a public hotspot while on the move or through a subscriber-owned access point in a residential or enterprise home zone area) was not available, keeping subscribers connected most of the time through their primary infrastructure and using the expensive cellular network only on an as-needed basis.

ILECs or cable MSOs might consider approaching the issue in a slightly different way. They often find themselves in the uncomfortable situation of needing to fight VoIP newcomers and mobile operators along with their traditional competitors. However, they may be better positioned to compete in the FMC arena by taking advantage of their vast last-mile networks to round up their triple-play offerings with wireless.

The case of converged operators may also present unique challenges. When both mobile and fixed networks are owned by the same operator, its goal would be to attract as many subscribers as possible to sign up for packages consisting of both fixed and mobile services, and to retain existing customers by some service enhancement and upgrade path. FMC appears to be a perfect way to satisfy this objective, offering the cost savings associated with service bundling, the convenience of unified services, a single bill, and a single support number for both fixed and mobile services.

The strategies for such an operator should heavily rely on home-zone pricing, which should be associated with a strong business case, since the external partnerships are not needed. The market rollout may occur faster and can be accompanied by marketing campaigns stressing better service integration and more uniform quality in comparison with competition. For existing customers, the upgrade opportunities and information about the benefits of FMC would be delivered in classic ways, such as mailed info, for example, alongside the monthly bill.

Summary

In this chapter we have spent a great deal of time digging into the business of fixed-mobile convergence. We have outlined the ecosystem of users and service providers whose worlds are being radically altered by the introduction of this new telecommunications concept. Further, we have outlined the FMC business case in terms of its main drivers satisfying diverse needs of subscribers and service providers alike. We also introduced the main FMC drivers in the framework of the four *C*s of FMC, *cost, coverage, capacity,* and *convenience,* and discussed the FMC deployment and proliferation barriers and the ways to overcome them.

The discussion in this chapter has illustrated the FMC benefits for both fixed and mobile operators, allowing the former to extend their offerings with mobile service and the latter to address issues associated with today's monolithic cellular networks.

We are now ready to move to the discussion of technical aspects of FMC. The next three chapters decompose the subject of FMC into its constituent fixed and mobile communications technologies, describing the relevant aspects of each and rounding up with the discussion of convergence enablers.

3

The *F* in FMC

"...it takes all the running you can do, to keep in the same place."

—Lewis Carroll

Fixed and mobile convergence is a phenomenon that is affecting and will continue to affect many aspects of networking and service delivery. Potentially almost all of today's telecommunication services can converge in a single offering over any network. To build the necessary knowledge base to understand convergence, in this chapter we focus our attention on the way services are being delivered in *fixed* networks.

The State of Fixed Networking

Contrary to popular belief, perhaps inspired by the fast pace of the mobile communications industry growth experienced in the past few years in developed countries; the astonishing mobile subscriber base expansion in developing nations such as China, Africa, and South American countries; and the increasing rate of fixed-to-mobile substitution, the state of the wireline telecommunications industry nowadays is very far from stagnant. Today's fixed telecommunication is as exciting, fast moving, and innovative as ever, possibly even more so than wireless. Indeed, it is actively embracing innovation, often spearheading the adoption of technologies initially invented for wireless networks (such as the IP Multimedia Subsystem, or IMS, which is discussed in Chapter 5).

The support of voice or audio communications between humans or between machines and humans (as in the case of interactive voice response systems—IVRs) still represents the bulk of the revenue and profits generated by telecom operators, despite the constant margin erosion due to intense price competition and the growing consumption of other services, such as broadband Internet access and value-added services such as mobile broadcast or media distribution and sharing.

Our overview of tethered communication starts with a discussion of the history of the way voice has been supported in fixed networks; it then moves on to a description of its latest incarnation: modern VoIP architectures.

Traditional Voice

Telephony came about as the commercial application of the invention by Alexander Graham Bell (or Antonio Meucci, depending on the reference source and point of view on who was first). The early wireline voice interconnection or switching capability was provided by the graceful efforts of thousands of switchboard operators (incidentally mostly female) who manually interconnected wires at the premises where the switchboards were hosted.[1] The first commercial telephone exchange began operations on January 28, 1878, in New Haven, Connecticut. The District Telephone Company of New Haven started up with eight lines and twenty-one subscribers, who paid $1.50 per month. By February 21, 1878, when the company published the first telephone directory, fifty subscribers were listed (quite an increase for one month of operation!).

This growth was so phenomenal that the number of telephone operators and the cost of the real estate required to handle the telephone traffic were soon no longer sustainable by telephone companies, which prompted the introduction of the automated system. The first patent for an automatic telephone switch was granted in 1879, but the first electromechanical telephone switch was invented and put into operation in 1891 by Almon B. Strowger in Kansas City (U.S.A.).

His invention was not exactly driven by the need to replace humans with automatic devices for cost or real estate savings, but rather by the need to protect himself from manual operators' potential (and actual!) malicious behavior! Mr. Strowger was not entirely happy with the local telephone operators intentionally diverting the calls from his potential customers to a rival company. It was perhaps the first example of the reason why telecom later became a highly regulated industry (even though humans are no longer directly involved in call switching).

From those days to the 1980s, when digitalization of the telephone network happened following the invention of the transistor and the subsequent ever-decreasing cost and ever-increasing power of computing, the telephone switches and network (also known as PSTN, or Public Switched Telephone Network) continued to evolve rapidly but maintained pretty much unchanged architectural principles.

Those included in-band signaling (that is, transmitting signaling used to select call destinations within the same frequencies used for speech) and analog voice. In-band signaling was implemented in the form of multifrequency tones (used in trunking) and dual-tone multifrequency (DTMF—still used in user-to-network signaling). Only toward the 1980s did the industry shift to digital communication between switches.

---

[1] The terms *tip* and *ring*, referring to the two copper wires (electrically positive and negative, respectively) carrying telephone conversations, and so familiar to today's central office and outside plant engineers, are rooted in these early manual exchanges, where they were used to describe the elements of a phone plug.

The new technology was called *common channel interoffice signaling,* which was the predecessor of Signaling System #7 (SS7), currently in use in digital telephone switches.

Digital telephone switches also gradually introduced the transport of voice as a digital signal and allowed not only for the separation of the control and media components involved in a telephone conversation, but also for more economical and efficient call switching, routing, and transport, necessary to handle ever-increasing amounts of traffic. The telephone network then became a hierarchy of access nodes and switches carrying voice over TDM (time-division multiplexing) digital trunk links. Previously these links were structured according to a Plesiochronous Digital Hierarchy (PDH), but now they are more often based on SDH (Synchronous Digital Hierarchy) or SONET (Synchronous Optical Networking) technologies, discussed in more detail in the following sections.

The access to telephony services offered by the network was either analog, using normal phones with tones and DTMF-based access signaling, or digital, using Integrated Services Digital Network (ISDN) phones or analog phones attached to an ISDN terminal adaptor. However, the majority of customers kept on using analog telephony services in most countries, as there was no apparent justification to migrate to ISDN[2] (especially when DSL access became available, convincing telecom operators to concede that ISDN growth was over). So, the dream of an Integrated Services Digital Network delivering voice and data using a single circuit with the 2B+D channels started at this point to give in to the IP-based networks and Internet reality.

Once voice was carried in digital format over the trunks between switches, it became very tempting to start considering it merely a different kind of data. This in turn opened the path to the introduction of the transport of voice over a packet data network, to benefit from statistical multiplexing. Originally, packetized voice transport was thought to be based on Asynchronous Transfer Mode (ATM)—specifically developed to carry delay-sensitive traffic such as real-time voice and video. But then the overwhelming success of the Internet made the Internet Protocol (IP) more compelling, making Voice over IP (VoIP) a reality. This shift was not only driven by the evolution of voice media transmission, but also by the explosion of VoIP as an end-user service in the enterprise and residential environments.

The following two subsections look at some technical details of the PSTN technology. Readers more interested in delving into VoIP right away may "fast-forward" to the later section "Voice over IP."

Signaling and Media   In order to set up, tear down, and control a voice communication between two endpoints, it is necessary to have an end-to-end voice communication path set up through the telephone network. This means that either there is an agreed end-to-end way to signal the establishment of the communication path or that along the way it is possible to perform an interworking between different signaling methods. Not only that: If the end-to-end path between the parties in a voice communication includes segments

---

[2] Often characterized by the phrase "too little too late" or mockingly referred to as "innovation subscribers didn't need" (ISDN).

where voice is carried as an analog signal and segments where voice is represented in digital format, there needs to be encoding/decoding at the boundaries of the segments.

The downside of this encoding/decoding, if applied multiple times, is that it introduces jitter and processing delay, which decrease the overall quality of a voice communication. So it became clear that it was beneficial for the industry to move quickly and in a coordinated fashion to evolve the telephone network from a system based on in-band signaling and analog voice transmission toward a system transmitting media and signaling over a digital network. Such a network would be based on computers exchanging signaling over data links to set up digital voice transmission circuits between digital switches connected by TDM trunks.

If this was not well planned, an end-to-end communication path would have required too many signaling and voice-encoding interworking steps, and the quality of user experience the whole system delivers would have substantially suffered. For this reason, the role played by standard organizations and regulatory bodies such as International Telecommunications Union–Telecommunication Standardization Sector (ITU-T)—better known at those times as CCITT, standing for Comité Consultatif International Téléphonique et Télégraphique—was invaluable, as they provided the basis for interoperability, and the governments or regional agencies could enforce the timeline for deployment of new standards cross-border or inside a country. However, international standards for signaling were also specified, and national regulations were defining country-specific standards or variants of telecommunications standards for signaling, which required some translation at international traffic exchange points. Similarly, different voice-coding standards were specified by the ITU-T, which also required transcoding.

Having provided this brief historical perspective, we will now focus the discussion on the most recent signaling and voice-encoding methods used in the PSTN, forgoing a lengthier discussion on all signaling methods used in the history of telephony, as these, albeit interesting, will not be used in the FMC discussion to follow. Whenever appropriate, some reference to historical aspects related to PSTN operation will still be made, though.

Signaling   The signaling system that allows the PSTN to operate is like the nervous system of the human body. It allows the exchange of information between the periphery of the network and its control centers, and among the control centers themselves to coordinate the operation of the network and the setup of end-to-end calls, much the same way as the nervous system can coordinate muscles and other organs and senses to allow life forms to exist.

The PSTN operates using a *mix* of signaling techniques. Analog techniques are normally used for communication between analog phones and the network (ringing, dial-tone, and user-to-network signaling based on DTMF tones).[3] In-band digital signaling

---

[3] Those tones we well recognize when we push keys on the telephones' keypads, or that we hear when selecting one of the options in a menu offered by a typical IVR of an automated airline or car rental customer services center.

techniques are used between digital devices, capable of generating, transmitting, receiving, and processing digital information. Examples of these are primary-rate ISDN access (PRA)—defined in ITU-T Recommendation I.421 [5]—or basic-rate ISDN access (BRA)—defined in ITU-T Recommendation I.420 [4].

Out-of-band (also known as common channel) digital signaling is normally used between PSTN network nodes, also known as switches. The Signaling System #7 (SS7) standard by the ITU-T defines an out-of-band signaling system. Figure 3.1 shows the SS7 protocol stack.

MTP (Message Transfer Part) layers of the SS7 protocol stack, including layers between Physical and Network, handle transport-related aspects of signaling traffic. SCCP (Signaling Connection Control Protocol) creates reliable transport connections; it operates at the Transport/Session layers of the ISO-OSI model. TCAP (Transaction Capabilities Application Part) manages signaling transactions. The various user parts and application parts are devoted to signaling network applications. All application parts are naturally operating at the Application layer.

SS7 messages are originated and terminated (or routed) by nodes in the signaling network. The system consisting of such nodes provides the basis for both basic call setup services (the ISDN User Part—ISUP—and the Telephony User Part—TUP [58] [59] [60] [61] [62]—do exactly that) and intelligent control functions. Such functions include Intelligent Network (IN)–supported services (through the Intelligent Network Application Part, or INAP [15]), and mobile telephony (through the Mobile Application Part (MAP) [16] in GSM systems and ANSI.41 in CDMA and North American TDMA and CDMA systems).

The ISUP standards [17] [18] [19] [20] [21] define the procedures and protocol-level details of the signaling application used between telephony switches to set up, manage, and tear down trunk circuits that carry voice and data calls over the PSTN. Since ISUP manages circuits over inter-switch trunks, it is used for both ISDN and non-ISDN calls, despite its name, which may suggest otherwise. Calls that originate and terminate at

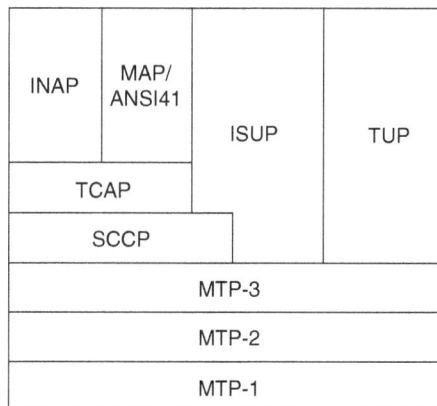

Figure 3.1   The SS #7 protocol stack

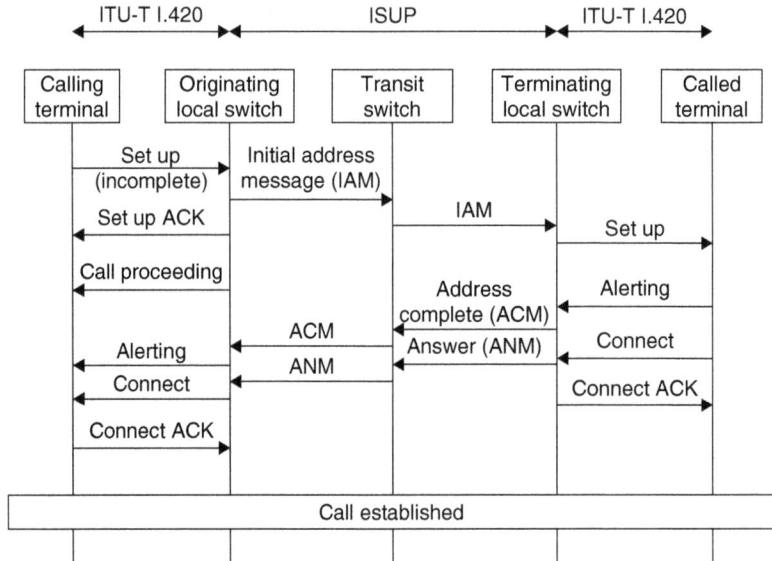

Figure 3.2   Relationship between ISDN signaling and ISUP

the same switch do not use ISUP signaling.[4] Figure 3.2 shows how ISDN basic rate access signaling defined in I.420 [4] and ISUP relate to each other.

The INAP layer in the SS7 model is used to define advanced telephony services by enabling interaction between service switching points (SSPs) and service control points (SCPs) in the network. SCPs implement service logic that is invoked by defining some triggers in the network. Triggers are "armed" at telephone switches and cause switches—acting as SSPs—to initiate an "Intelligent Network" transaction.

For instance, it is possible to define premium rate services by allocating ranges of numbers to specific functions and tariffs, and it is also possible to implement smart call routing depending on certain parameters such as time of day (e.g., calls may be routed to U.S. call centers in daytime and to India during the night) or to deliver services such as tele-voting.

Other SS7 protocols such as MAP and IS-41 enable interaction with location registers in GSM and CDMA, respectively. Those protocols have important roles in support of cellular wireless macromobility and therefore take center stage in the FMC discussion to follow in Chapters 4 and 5.

---

[4] For the records, there was also user-specific signaling called TUP (Telephony User Part) [58] [59] [60] [61] [62]. In addition, some countries also used country-specific forms of this signaling (for example, the National User Part adopted in Great Britain).

When ISDN is not offered to customers, analog signaling in the access part of the network supports dial, ringing, busy, and alerting tones and not only communicates to the network the destination number a caller would like to reach, but also selects and invokes a set of telephone network services, or *supplementary* services,[5] the PSTN offers to end users. Examples of common supplementary services include:

- *call forwarding unconditional* (allows a redirection of all incoming calls to another number)
- *call waiting* and *message waiting indicator (MWI)* services
- Additional services such as conditional call forwarding or even wake-up calls!

This signaling is based on DTMF tones for analog phones.

In fact, in analog telephony the telephone is a passive element in the service provisioning process, essentially translating user input through the keypad into DTMF signals toward the network, which interprets these signals and processes the service logic. In contrast, when digital signaling is used between the telephone and the network, as in the ISDN, the telephone becomes a digital processing–capable device.

In networks with digital signaling support, then, two different, supplementary, service delivery implementation options exist. One option is known as *stimulus-based,* and really this replicates the model of a "dumb" analog phone used as a DTMF-relaying slave device. This option may support the use of some special "feature keys" on a telephone and the transmission of the pressed feature key number to the network via the I.420 "INFORMATION" message. In this case the network needs to keep a map of what telephone a customer is using and what feature key is mapped to what service.

The other option is known as *functional,* as the telephone is aware of the selection of a service and cooperates with the network in the delivery of the service itself. This method uses the I.420 "FACILITY" message if the service is associated with a call establishment or clearing and the "REGISTER" message if the service can be invoked independently of these events. Special messages have also been defined, such as a message to put a call on hold.

Media   The PSTN uses 64 Kbps digital circuit-switched channels carrying PCM-encoded (pulse coded modulation) speech. PCM modulation is based on sampling a voice signal (after it passes through a low-pass filter at 3.4 KHz) 8,000 times per second and representing the voice sample by means of 8 bits. The encoding process takes place at the point of termination of the local loop as described in the next section, which is devoted to the description of the PSTN components.

Without going into deeper discussion on how voice encoding is performed and on the different digital compression laws adopted in various regions of the world to represent the voice sample in 8 bits, suffice it to say that the unified encoding standard adopted in

---

[5] Also called *supplemental* services in some sources. Both notations are correct; however, we are going to stick to one for consistency.

the circuit-switched PSTN for digital voice is globally known as ITU-T G.711 [6]. Other voice-encoding standards are adopted to compress voice to lower rates (like adaptive PCM, G.722 [22], and others).

Technology Components   Much in the same way as a human body is made of different organs, and it works perfectly (most of the time) if all of them (bones, muscles, nerves, sensorial organs, etc.) are in good order and do not fail, a modern circuit-switched telephone network consists of a set of components that connect together and cooperate to deliver the end-to-end voice communication services. These include

- End-user devices
- Access network (also known as the local loop)
- User line termination (systems terminating customer physical lines such as wires or fiber at local exchanges)
- Local switching subsystem
- Trunking subsystem
- Signaling network
- Numbering and addressing system.

In the next few paragraphs we are going to review these components involved in delivering voice services in the PSTN.

The End-User Device   End-user devices in the PSTN include analog and ISDN phones. The "classic" fixed-line analog or ISDN phone has the well-known shape with a handset and a base where the "hook" and keypad reside (including possible feature keys, used, for instance, to access network-hosted voice-mail services or supplementary services), but more creative shapes have come to market recently, as shown here.

Also, in recent years cordless phones have become increasingly popular. Cordless phones today are mostly digital, such as those based on DECT—a digital cordless telephony standard discussed in Chapter 1—or a proprietary digital technology, but in the past they also were analog, similar to the first generation of cellular phones. As mentioned in Chapter 2, the form factor of a cordless phone was normally bulkier than that of a GSM or CDMA cellular phone, but more recently the trend is to produce models gradually starting to resemble cellular phones in their shape and user interface (if not the size).

The Local Loop   The *local loop* is the part of the fixed telephone network that grooms and distributes media and signaling between the customer premises and a central office (CO), where the local circuit switch is installed. It is used by all services such as plain old telephone service (POTS, a way to refer to classic analog telephony). The local loop is also popularly referred to as the "last mile," in that this is the last hop between the telephone network and a customer, implemented over copper twisted pair (or, more recently, fiber strands), most of the time spanning relatively short distances of just a few miles. The collection of local loops is known also as *outside plant*.

Due to the labor-intensive deployment and maintenance of the wired lines and the cost of the cables and facilities deployed (such as telephone poles, underground conduit, and tons of copper for the cables themselves), it constitutes the major investments and expenditures (in terms of both CAPEX and OPEX) undertaken by incumbent telecom operators, and that is why in many regions of the world they have been forced by regulators to open up the local loop to competitors (under the terms of "local loop unbundling" agreements) to presumably[6] foster competition and benefit end users. For this reason, also, the technology called *wireless local loop* has been deployed in countries or regions that could not afford the magnitude of investment required to support a wired local loop (especially in remote rural areas).

The Local Switching Subsystem and Subscriber Line Termination   Local switches (also known as *end offices* or as *central office (CO) switches* in North America or Digital Local Exchanges [DLEs] in the UK, just to name a few examples) terminate the local loop and switch traffic between the user premises directly attached to the switch or concentrate traffic onto trunking lines toward remote local switches or intermediate transit switching nodes. Terminating the local loop implies transforming analog voice signals from analog telephones into digital format, generating signals such as dial tone and ringing, and implementing services such as supplementary services. Subscriber line management; line supervision; and other operation, administration, management, and provisioning (OAM&P) functions are also located at the local switch.

Sometimes, "remote switch units" are deployed to economize on the transmission lines required to connect customers to switches at exchange offices (common in non–densely populated areas). In these cases, the subscribers handling functions are located

---

[6] The downside of such legislature is that to maximize profits in this newly created competitive environment, operators are often forced to cut outside plant maintenance and investment budgets, resulting in lowered standards and subsequently worsened overall quality of the network and service.

remotely from the switch, and only switching functions on TDM trunks, carried by feeder links between remote units and the switch, are performed at the local switch.

The Trunking Subsystem   Transport links between switches are defined as a hierarchy of digital TDM trunks of various capacities. These hierarchies have been defined differently in different parts of the world (E hierarchy in Europe, T [or DS, for Digital Subscriber] hierarchy in North America, and J hierarchy in Japan), as described in the Table 3.1.

More recently, synchronous transport hierarchies have been defined, specialized for optical transmission (which allows reaching data rates, capacities, and transmission distances over optical fiber facilities not possible with copper transmission lines).

There are two synchronous transport hierarchies:

- SDH (Synchronous Digital Hierarchy) developed by ITU-T [8]
- SONET (Synchronous Optical Networking) [9] developed by the ANSI T1 Committee.

The SONET hierarchy is mostly used in North America, the SDH hierarchy in the rest of the world. These hierarchies are defined in Table 3.2.

These transport links hierarchies are used to interconnect local switches, potentially via *tandem* (also known as *transit*) switches. There can also be a hierarchy of tandem switches, each switching trunks of higher capacity between regions of a national network or at the boundary between national networks. These trunks were normally second-choice trunks when direct (high-usage) trunks existed between nodes

TABLE 3.1   Plesiochronous Digital Hierarchies

European (E)	North American (T/DS)	Japan (J)
64 Kbit/s 1 channel	DS0 64 Kbit/s 1 channel	64 Kbit/s 1 channel
E1 – 2.048 Mbit/s 32 channels	T1/DS1 – 1.544 Mbit/s 24 channels	J1 – 1.544 Mbit/s 24 channels
–	DS1C 3.152 Mbit/s 48 channels	–
E2 – 8.448 Mbit/s 128 channels	T2/DS2 – 6.312 Mbit/s 96 channels	J2 – 6.312 Mbit/s 96 channels or 7.786 Mbit/s 120 channels
E3 – 34.368 Mbit/s 512 channels	T3/DS3 – 44.736 Mbit/s 672 channels	J3 – 32.064 Mbit/s 480 channels
E4 – 139.264 Mbit/s 2048 channels	T4/DS4 – 274.176 Mbit/s 4032 channels	J4 – 97.728 Mbit/s 1440 channels
E5 – 565.148 Mbit/s 8192 channels	T5/DS5 – 400.352 Mbit/s 5760 channels	J5 – 565.148 Mbit/s 8192 channels

TABLE 3.2   SDH and SONET Hierarchies

Line Rate (Mbit/s)	SONET Level/ Frame Format	SDH Level and Frame Format
51. 840	STS-1/ OC1	STM-0
155. 520	STS-3/ OC3	STM-1
622. 080	STS-12 / OC12	STM-4
1,244. 160	STS-24 / OC24	STM-8
2,488. 320	STS-48 / OC48	STM-16
4,976. 640	STS-96 / OC96	STM-32
9,953. 280	STS-192 / OC192	STM-64
39,813. 120	STS-768/ OC768	STM-256

lower in the hierarchy. In North America, for instance, a hierarchy of tandem switches had been defined as follows [7]:

- **Class 1**   Regional Center, used for international gateway functions.

- **Class 2**   Sectional Center, fully meshed to all other Class 2 centers, mostly used as the last-resort route of a call.

- **Class 3**   Primary Center, may not be directly connected to all Class 3 tandem switches, but go through another Class 3 or a Class 2 node to reach other Class 3 nodes.

- **Class 4**   Toll Center, handling traffic in and out of local switches (Class 5); this class was replaced by Access Tandems at the time of the divestiture.

For those wondering where "Class 5" is in this list, recall that a Class 5 switch is a local switch and would mostly terminate customer lines (even though it can perform the trunking role for remote switch units). So, it should not be mentioned in the trunking subsystem, other than to note that it is at the edge of it.

In more recent times, the number of hierarchical levels in the telephone network has been shrinking as higher- and higher-capacity links are being carried on packet infrastructures, or over SDH optical transmission infrastructures, allowing more direct connectivity between nodes.

The Signaling Network   SS7 messages are originated, terminated, and routed by the nodes belonging to the signal network. When a switch hosts the capability to invoke IN services, it is called the service switching point (SSP). The SS7 nodes include non-IN-capable switches, SSPs, service control points (SCPs), and the signaling transfer points (STPs). Signaling can originate in an SSP or the SCP, and the originating and terminating addresses in the SS7 messages are called, respectively, the Originating Point Code and the Terminating Point Code. The SCP hosts application's service logic, which it invokes according to a specific request from the SSP. SCPs may implement special charging or call routing rules (like those needed for toll-free service, premium rates, etc.).

The STP is a node in the signaling network used as a signaling router. STPs can also act as gateways between signaling networks. In this case, they could also act as signaling protocol translation entities (e.g., entities that translate among different national and international formats of signaling protocols). STPs are normally deployed in mated pairs to provide redundancy in support of fault tolerance. SS7 messages are sent normally by their source at a particular hop in the end-to-end path toward both STPs in a mated pair, and an SS7 protocol stack can manage the reconciliation of duplicate messages. The following illustration provides a view of a classic SS7 network:

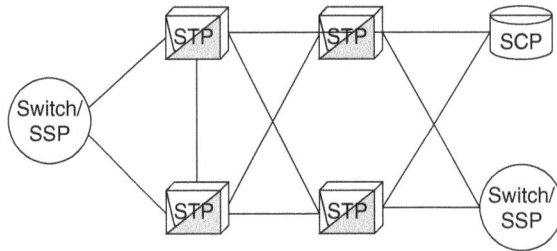

Numbering and Addressing   In order to place a phone call or access a telephone network service, we normally dial a sequence of digits commonly known as a "telephone number." The numbering of telephone networks is defined in ITU-T Recommendations E.164 [10] and E.123 [11]. E.164 defines the format of telephone numbers as well as the criteria for the reservation, assignment, and reclamation of numbers. E.164-compliant numbers have a maximum of 15 digits. It should be noted that this number of digits does not include the international prefix we normally have to dial on the keypad of a telephone when we want to place an international call. E.123 describes how to represent an international telephone, starting with a plus sign (+) and the country code.

Voice over IP

Having discussed the technology components of circuit-switched telephony, we can now move on to consider Voice over IP, fast emerging as the next-generation technology capable of replacing POTS and redefining PSTN as we know it.

The massive uptake in broadband access services in the residential market and the widespread migration of the corporate PBX infrastructure to IP-based systems driven by cost and functionality benefits underscore the emergence of VoIP technology in wireline networking. The access to fixed IP networks via Wi-Fi in enterprises, institutions such as hospitals and colleges, and residences has furthermore permitted the expansion of VoIP into the wireless domain, specifically Wi-Fi transport (also called VoWi-Fi).[7]

In this section we describe the general technology aspects of VoIP, which are largely applicable to wireline and wireless access alike, with the initial focus on the wireline networks.

---

[7] Support of VoIP in cellular wireless networks in the not-so-distant future is definitely in the cards as major mobile operators around the world are entering advanced stages of IMS trials.

The introduction of VoIP is quite a revolutionary concept in the way voice communication occurs. The Internet is an open network providing end-to-end network layer connectivity between Internet hosts. Over the Internet, there is in principle no need to use standard signaling and media encoding to establish a voice communication between Internet hosts. Any two users may agree on their own signaling protocol to set up the communication session, choose their own voice-encoding method, and build their own piece of software implementing these agreements. Provided that the end-to-end communication path offers adequate quality of service, the two users may use their own mechanism to communicate directly, regardless of distance, location, and time of day, as long as they have access to the network.

This theoretical exercise is useful in understanding the dramatic effect of the end-to-end network layer connectivity offered by the Internet in support of any service, including voice communications, in the telecom industry. Any group of Internet users may in fact agree on their own way to define voice communication service. It is, therefore, possible for a venture to become a global player in offering voice communication services to a population of users who accept its terms and conditions of usage and service framework, as we are witnessing these days with Skype and other, similar players. It is not even necessary for such a venture to own an access network or even worry about the specific user access mechanism—cable or DSL line—so there are few or no technical barriers to offering a commercial VoIP service.

Of course end-user inertia or network externalities[8] may limit the uptake of VoIP services, so it is fundamental for VoIP service providers to offer connectivity toward the legacy PSTN network (for example, SkypeOut service). While high-quality voice communication over the Internet today is possible, interworking with the PSTN will remain a necessity and the PSTN modus operandi will continue to influence and shape the service offerings of the VoIP service providers (at the very least for those aspects related to service connectivity with the PSTN itself).

It is, then, no wonder that in the end, to realize the economies of scale and make use of equipment designed for interoperability with other standard VoIP solutions as well as the PSTN, many VoIP service providers feel compelled to adopt standard signaling protocols and media-encoding formats for an optimal operation of their network. Typically, this is the case for traditional service providers venturing into the field of VoIP. We will see in Chapter 5 how this approach especially makes sense for offering services with a common look and feel over any access technology.

Indeed, standard signaling methods between the terminal and the network are practically required when the user terminal is no longer a computer but a handset compatible with many service provider infrastructures and capable of connecting to multiple networks while roaming.

Signaling and Media   As there are drivers to deploy VoIP in many different ways to satisfy different user needs, it can be difficult to define a dominant VoIP signaling or

---

[8] Network externality is the reluctance of users to join a network of users of a given telecommunication service if they can only communicate to users of that network, as they may not be able to communicate with people who have not yet joined that network.

encoding method. However, for the scope of this book, we shall focus mainly on the signaling and media transport protocols used in the IETF, ITU-T, and 3GPP standards such as the Session Initiation Protocol (SIP) [23], MEGACO [24]/H.248 [25], DIAMETER [26], and the Real-Time Transport Protocol (RTP [27]). Unlike in our circuit-switched telephony discussion, we shall not focus on media encoding itself, since in IP networks any media encoding is possible as long as a certain protocol such as RTP has been defined to support its transport.

Signaling   We will start our discussion of signaling with the description of the Session Initiation Protocol. SIP is a text-based protocol that can be carried over UDP or TCP. At a high level, SIP can be used in VoIP to do the work of both SS7 and POTS or ISDN access signaling. SIP is used to signal the establishment of a multimedia session among IP endpoints using a client/server model. SIP endpoints are called user agents (UAs). Thus the user agent client (UAC) is the calling party, and the user agent server (UAS) is the called party.

SIP is not limited to establishing a signaling path for voice communications and can in fact be used to set up video, text (for instant messaging), and other types of media sessions. The exact media component(s) in the multimedia session are described using a Session Description Protocol (SDP) [28], which provides the two endpoints with information about media, such as:

- The format of the media (G.711, AMR, etc.)
- The type of media (video, audio, etc.)
- The transport protocol (RTP, etc.)

For an IP multicast session, the following information is also included:

- Multicast IP address for media
- Transport port for media

For an IP unicast session, the following information is also included:

- IP address for media
- Transport port for contact address

It should be noted that SDP was originally conceived to be used in conjunction with the Session Announcement Protocol (SAP [29]) to advertise multicast sessions on the Multicast Backbone (MBONE—an experimental multicast distribution network mostly used in academic or research environments in the 1990s). SDP then was carried in the payload of a SAP packet to describe multicast sessions available on the MBONE and used to describe each entry of a sort of a TV program guide. Much in a similar way, to establish a multimedia session, SDP is now carried inside the headers of some SIP messages and describes the way to handle media components of the SIP session.

SIP works using requests and responses similar to the way other protocols such as HTTP (the Hypertext Transfer Protocol) are used to access Web content. Each SIP signaling transaction consists of a request that invokes a particular method, or function, on the server and at least one response. SIP users are identified in a way similar to e-mail users, via a URI (Uniform Resource Identifier). A SIP URI is an identifier (such as telephone number or a username) accompanied by a domain identifier, and therefore normally appears as follows:

```
sip:123456781@ACME.com
sip:alice@alicedomain.com
```

There also exists a specific form of SIP user identity called "tel URI," defined in IETF RFC 3966 [30], to represent telephone numbers, which can take any of the following forms:

```
tel:+1-201-555-0123
tel:7042;phone-context=example.com
```

When a phone-context is specified in a tel URI, the number is unique and routable within the specified context (the domain specified).

Using these formats, the Domain Name System (DNS) can therefore be used to resolve a username to an IP address of the user, much in the same way as an e-mail address can be resolved to the IP address of a mail server, which can handle e-mail for the user. A SIP message is also somewhat similar in format to an e-mail message, since it is also text based, and it uses the Multipurpose Internet Mail Extension (MIME) [31] to describe the content. SIP messages can therefore contain information such as graphics, authentication tokens, or video, which could be useful in supporting advanced features such as carrying images of the calling party. More importantly, it is possible to carry binary information such as ISUP binary objects within a SIP message, when preservation of ISUP data is necessary for the proper operation of PSTN features implemented using SIP [32].

SIP's text-based nature, which is quite common in many IETF protocols, has both its positives and negatives. While this format helps in troubleshooting without the aid of special conversion tools, it has exposed SIP to criticism, as encoding of information as text is sometimes lengthy, inefficient, and not always suitable for some applications (especially in the wireless environment or in conditions where the length of messages needs to be minimized to attain some desirable level of performance or fit into wireless technology–specific Layer 2 frames).[9]

---

[9] Quoting Igor Faynberg, a Bell Labs telecom visionary, and chair of the PSTN/Internet Interfaces (PINT) IETF working group: "Had SIP been binary-encoded, a simple tool could translate it into the text (and back) for troubleshooting as is done with all other protocols. . . . The decision to make SIP text-based was based only on the Internet habit (actually, a religion) to make all application protocols text-based. But there is actually a subtle more serious reason: If SIP were to be binary-encoded, it would have to use the ASN.1, which at the moment of SIP creation was not ready to support an efficient encoding that it can support now."

SIP is still undergoing evolution. SIP messages have been defined and enhanced over time as the functionality of SIP-based solutions evolved to meet specific user requirements. In fact, SIP was initially a very simple protocol that was not on a par with ITU-T Recommendation H.323 [33] (an ITU-T-defined protocol for VoIP sessions setup and management, offering compatibility with ISDN access protocol), so it had to be extended to offer previously not supported telephony features. Also, SIP was greatly enriched to support important functionality, often extending beyond signaling for VoIP applications, such as presence, conferencing, ability to carry text messages (e.g., the SIP MESSAGE method) and to interact with network resources, admission control, and management subsystem.

It is not the goal of this book to get into deeper details of SIP, as this technology is widely covered in today's publications. For those readers wishing to probe further, we invite them (no pun intended, as "INVITE" happens to be the name of one of the SIP methods) to explore some of the material listed in the bibliography at the end of this book or to access directly the RFCs on the IETF Web site. Table 3.3 is largely limited to a description of SIP methods directly related to the main subject of this book and the convergence discussion to follow.

TABLE 3.3   SIP Methods Description

Method	Method Usage
REGISTER	A UAC sends a message to inform a UAS of the IP address where it is currently reachable.
The INVITE	Requests an endpoint to join a SIP session, or, during a call, it requests an endpoint to change session parameters.
ACK	Sent in response to an INVITE.
OPTIONS	Queries the capabilities of a UA.
CANCEL	Ends a session, which could not be fully established.
BYE	Ends a session, which was previously successfully established.
INFO	Defined in RFC 2976 [34], this is used to allow for the carrying of session-related control information that is generated during a session, such as ISUP and ISDN signaling messages used to control telephony services between media gateways or between an ISDN IP phone and the network. This may also carry billing data such as advice of charge.
PRACK	Defined in RFC 3262 [35], this is a provisional acknowledgment sent in a reliable manner (it was needed because provisional responses sent unreliably are not suitable in a number of telephony and PSTN interworking applications).
UPDATE	Defined in RFC 3311 [36], this allows a client to update parameters of a session similarly to a re-INVITE, but it can be sent before the initial INVITE has completed.
REFER	Redirects the sessions to another IP address.
SUBSCRIBE/NOTIFY	Defined in RFC 3265 [37], this allows a UA to subscribe to a SIP events package, so that it can be notified when a specific event happens via a NOTIFY method.

The SUBSCRIBE/NOTIFY methods along with the PUBLISH method [38] comprise the SIP extensions for Instant Messaging and Presence (SIMPLE). The term SIMPLE is used to refer to both the working group in the IETF that is producing specifications for presence and Instant Messaging (IM) within the SIP framework and the collection of protocol specifications produced by that working group. Chapter 6 provides further discussion on this interesting technology.

SIP entities can also send a variety of messages as a response:

- "100" responses define informational or provisional responses, such as 180 Alerting.
- "200" responses notify a request was successful (like 200 OK).
- "300" responses redirect to a different IP address.
- "400" responses indicate a request failure, such as 480 Temporarily Unavailable.
- "500" responses identify a server failure, such as 503 Service Unavailable.
- "600" responses identify failure, e.g., 603 Decline.

As in public VoIP networks, it is necessary to identify endpoints, to allow them to use IP telephony services, and to know at what IP address a VoIP user is currently located. To achieve this there is a need to enable VoIP user registration and authentication, for which SIP is perfectly suited. The SIP entities in the network capable of registering users are known as SIP *registrars* (and are sometimes also known as SIP *servers, proxies,* and other more- or less-precise names to identify SIP *user agents* that can handle SIP signaling related to many SIP endpoints and act as SIP signaling routers, keep call state, and also implement some service logic, invoke external application servers, or access subscriber databases).

The registrars may interact with subscriber data servers or Authentication, Authorization, and Accounting (AAA) servers via AAA protocols such as RADIUS or DIAMETER [26]. The interaction with the subscriber database is used to determine the right of a user to access the VoIP service and also to install any user-specific rules that may be used to redirect SIP signaling to specific application servers. Application servers are used to implement the service logic of services the user may need to access as part of the VoIP service subscription (or even to implement advanced VoIP applications such as Voice Call Continuity (VCC)–based dual-mode services, as discussed in Chapter 5).

Application servers may change the contents of SIP messages or even initiate third-party SIP calls. They may also interact with other subscriber data repositories to retrieve application-specific user data (again via AAA protocols or other means such as Extensible Markup Language [XML]–based interfaces) or implement interworking functions with legacy IN services. XML, incidentally, is a general-purpose language that supports many different kinds of applications. Its primary function is to help sharing of data across systems connected via the Internet. It is widely used in database access, and more and more in network management applications.

The next illustration shows typical SIP session setup and tear-down scenarios, where SIP UA and proxies are included.

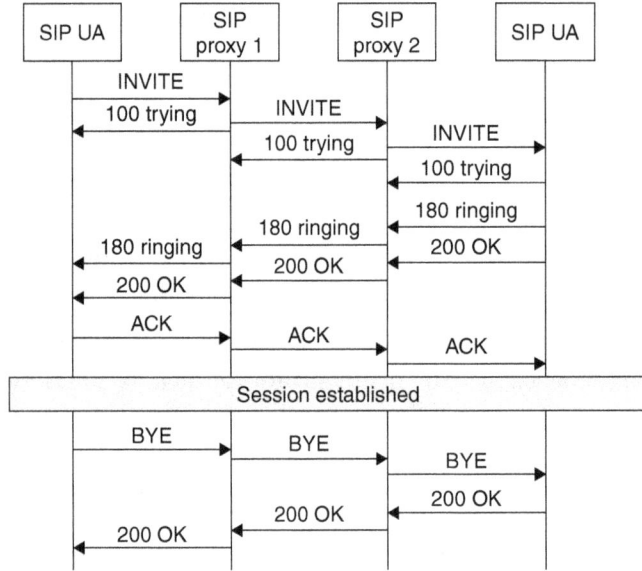

Media In IP networks, encoded voice[10] is normally carried using RTP, defined in IETF RFC 3550 [27]. RTP provides end-to-end transport services for real-time media, such as interactive audio and video. As IP networks do not guarantee in-order packet delivery, the sequence numbers included in the RTP header allow the receiver to reconstruct the sender's packet sending order. RTP also provides timestamps allowing the receiver to play back the real-time media it carries according to the sender's intended timing; however, RTP itself does not provide any mechanism to guarantee timely delivery. So, any QoS guarantees need to be provided by the underlying IP network carrying RTP payloads.

The RTP protocol suite also includes the RTP Control Protocol (RTCP) to monitor the quality of service experienced by receivers of RTP frames. This allows dynamic selection of the codec in a voice communication (for instance, the codec selected initially for a SIP session, such as a wideband codec, may provide the best quality of sound, but then the session may fall back to a narrowband codec once the detected QoS level is not sufficient to sustain the required bit rate).

Speaking of QoS, two main approaches are used to signal QoS requirements in an IP network: the Resource Reservation Protocol (RSVP [40]) and Differentiated Services [39] (DiffServ). The RSVP protocol reserves resources on a per-IP packet flow basis

---

[10] We again consider the case of voice in this section, in keeping with the spirit of the book's main subject, but video and other forms of media could also be part of a SIP-initiated session.

(e.g., identified by a combination of source and destination IP addresses, TCP port numbers, etc.) in the network by installing a soft state in the routers between the source and the destination (or two or more destinations, in case of a multicast session).

This soft state needs to be periodically refreshed by means of RSVP messages to keep it alive. This approach did not find wide acceptance in the marketplace, since the signaling of IP flow requirements in the end-to-end path proves not to scale well in the Internet core, where hundreds of millions of IP flows can sometimes be handled by a single router. Instead, an IP DiffServ approach has been adopted as the preferred way to identify the per-packet QoS treatment at every IP network hop. DiffServ does not attempt to reserve resources on a per-IP flow basis; it just handles traffic aggregates differently, depending on the marking of the Differentiated Services Code Point field of the IP header [50].

Media gateways, essential elements in the VoIP media-handling infrastructure, are used for interworking with the PSTN at the media level and to trunk traffic between circuit switches over IP or other packet technologies. They are also an essential part of the architecture of a new wave of PSTN switches based on a softswitch concept. Media gateways are in this case connected between themselves in trunking applications, and they would exchange control information, for example, using the Bearer Independent Call Control (BICC)–based signaling. In this case voice frames within the core network are carried over IP to benefit from transport efficiency in the backbone, but users would still be offered a POTS or ISDN service, not a VoIP service.

BICC [12] [13] is a signaling protocol specified by the ITU-T, based on narrowband ISUP, used to support narrowband ISDN service independently of media and signaling transport technologies. ISUP messages carry both call control–related and bearer control–related information, identifying the physical bearer circuit by a Circuit Identification Code, which is specific to circuit-switched TDM networks. In contrast to ISUP, BICC signaling is interoperable with any type of bearer, such as ATM bearers or IP networks, in addition to TDM. This allows separation of the signaling plane from the media transport plane. More recently, other options to exchange signaling between gateways based on SIP have been proposed in 3GPP (such as approaches based on SIP-I, also used to interconnect fully SIP-based IP networks).

Finally, to generate tones and announcements, the network may include Media Resources Functions (MRFs), which may be controlled directly via SIP signaling or via MEGACO [24] / H.248 [25].

VoIP Technology Components   VoIP could at least in theory be viewed operating as a simpler "organism" than traditional circuit voice systems. In fact, when VoIP technology is used for peer-to-peer communication (as opposed to the client/server model supported by service providers), there is no need to introduce any hardware in the network in addition to a user terminal capable of IP termination and supporting an audio interface to reproduce sound and record speech (for example, a headset and microphone).

Unlike with analog voice, VoIP can be supported by an appropriate software client running call state control, session setup signaling, and media encoding and decoding. Such a setup does not need to be standardized (with the exception of the IP network layer protocols, which are indeed standard), as we have discussed earlier in the chapter.

However, the complexity of this simple setup should not be underestimated, as what looks apparently simple is much more complex than common wireline phones and the relatively straightforward circuit-switched technology connecting them.

In commercial public networks offering IP telephony services, the scenario gets much more complex and many more entities are required. The client device is still required, but the network must now include elements (or functions) such as SIP proxies and servers, subscriber databases, an admission control subsystem (to allow users access to network resources based on predetermined policies), an interworking subsystem, and an address resolution subsystem.

Client Devices   VoIP client devices can support complex functionality, while at the same time having the familiar look and feel of a traditional POTS telephone, and are normally equipped with some feature keys, especially in the case of enterprise deployments. There are examples of cordless IP phones using Wi-Fi or Bluetooth as the wireless connectivity technology. Very frequently, though, the IP telephony service is accessed via *softphones,* that is, a piece of software running on a general computing platform and offering a "keypad-like" look and feel on-screen. Another relatively popular option is the use of USB (Universal Serial Bus) handsets, which are handsets directly connected to a computer via a USB interface, to control a softphone application running on it. These phones look very similar to a cordless phone and are sometimes used as a way to use services such as Skype, for instance.

Servers and Proxies   The SIP protocol operation in public networks requires a number of functional components. A SIP Proxy server, an essential component of any SIP network, is needed to perform call routing, user authentication, and authorization. A SIP Proxy server can forward the messages from a UAC to the next hop based on local routing tables or on a query of some routing database or another server (like a Redirect server); it then forwards responses from remote entities back to the UAC. Proxies may withdraw from the signaling path after the SIP session has been established, or, based on operator policy, stay in the signaling path (e.g., for billing purposes). UA or proxies can contact redirect servers to locate an endpoint. A redirect server may return a single IP address or multiple IP addresses.

When multiple IP addresses are returned, then the SIP message can be sent by the UA or proxy to all of them in sequence or in parallel (performing what is known as "SIP forking") to implement a "Follow Me" kind of service.[11] This function may or may not be collocated with the SIP Registrar server function, which allows a SIP UA to register its current location (e.g., the contact SIP proxy to forward a SIP message to, or an IP address) via a SIP REGISTER message. The SIP Registrar may also perform user authentication and authorization, and respond to location queries from other SIP servers (acting as Location Servers).

---

[11] When the call placed to a VoIP number is automatically forked to ring multiple other numbers supplied by the subscriber, like an office or mobile number.

Back-to-back user agents (B2BUA, another common element in the SIP infrastructure) are used to act as a UA server and client at the same time toward different SIP endpoints, with the aim of possibly changing the content of SIP messages, for instance to implement call features. SIP application servers typically play this role.

Admission Control Subsystem   Since a number of users share public IP networks, SIP sessions compete to get access to the network resources open for use by both VoIP media and signaling. It should not be forgotten that IP networks will at the same time be used to run a variety of other data applications. Even in the case of VoIP being provided with strict priority treatment, there may be other services in the same queue, such as real-time video and other mission-critical applications. At times of congestion, then, it can be necessary to decide which calls to admit and which calls to shed.

This may also have public safety significance, therefore a Resource Admission Control Subsystem (RACS) must be implemented in a large-scale commercial VoIP deployment. This subsystem, when used for VoIP applications (it could be used also for other operator-controlled applications not entailing SIP and media components), provides a SIP session with access to IP network resources based on a variety of application-related policies and charging rules, determined by the service provider.

As such, this subsystem ties together the SIP servers' layer, subscriber policy repositories, and the network layer, to implement decisions on admission to network resources. The ITU-T and TISPAN are working on the definition of this subsystem for fixed networks. The IMS defined by 3GPP is in this case the SIP Service layer infrastructure, but other applications are not ruled out. This subsystem, clearly one of the most critical elements in a viable public IP telephony network, is conceptually represented in Figure 3.3, where the subsystem is acting as an admission entity both toward other operators' networks and toward the access network.

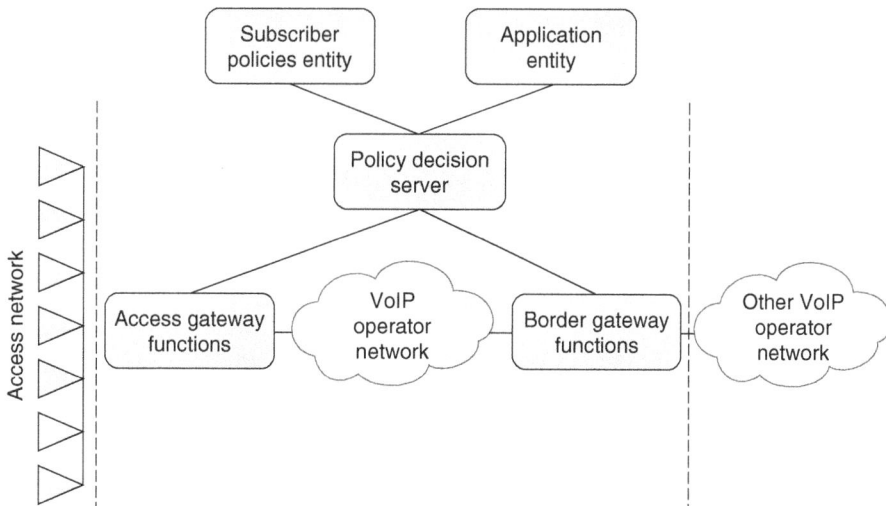

Figure 3.3   Conceptual representation of a Resource Admission Control Subsystem

Interworking Subsystem   The interworking subsystem provides users of a public VoIP network with connectivity to and from the PSTN and to other VoIP networks implementing differing signaling protocols (or even media encodings). This subsystem is also essential for the commercial success of VoIP deployments, as it offers access to the whole community of PSTN users, thus eliminating the problems of backward compatibility with legacy technology and the related network externality that may otherwise affect the widespread adoption of VoIP.

Entities typically involved in the interworking subsystems (as mentioned in the preceding section) are media gateways and media gateway controllers, also popularly referred to as *softswitches*. In typical implementations the media gateway control part is in fact very often split from the media handling part (performing transcoding, echo cancellation, etc.) to allow for independent scaling and procurement. The signaling protocol used between the control part and the media part of the media gateway has been defined jointly by the IETF as MEGACO – Gateway Control Protocol [24], and the ITU-T as H.248 [25].

The media gateway controller in a PSTN interworking scenario terminates BICC or ISUP signaling on the PSTN side and SIP signaling on the VoIP network side. The specification covering the aspects of SIP interworking with ISUP- or BICC-based networks is ITU-T Q.1912.5 [41]. In mobile environments, an equivalent specification exists defined by 3GPP (in 3GPP TS 29.163 [42]). A high-level model of SIP-to-PSTN interworking is depicted in Figure 3.4, where the CS network can be mobile or fixed and the PS network, hosting the SIP UA and the SIP proxy, can also be mobile or fixed.

It should also be noted that in the mobile environment, a specific form of SIP has been defined (the 3GPP SIP profile). Also, there is potential for IP version differences between mobile environments and fixed-line environments. 3GPP specifies the interworking

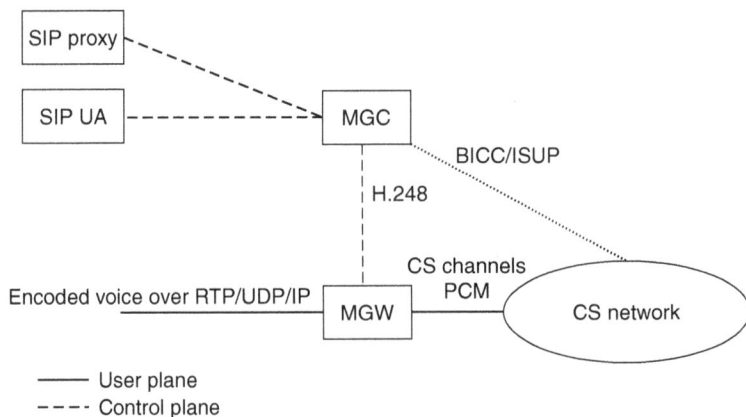

Figure 3.4 High-level interworking model 3GPP between a SIP-based network and a CS network (PSTN, cellular . . .)

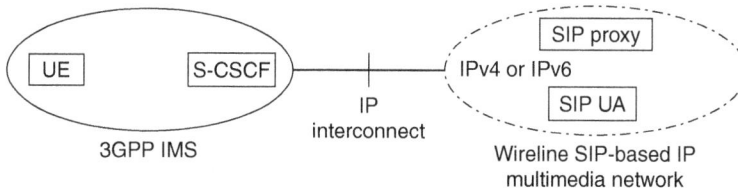

Figure 3.5   Interworking model 3GPP IMS with another SIP-based network as per 3GPP TS 29.162

between the IMS and other SIP networks, not adopting the 3GPP SIP profile, in 3GPP TS 29.162 [43]. Figure 3.5 represents the conceptual model of such an interworking method, where it is clarified that a SIP-profile interworking and perhaps an IP-version interworking between IPv4 and IPv6 may be required.

Other requirements may exist in order to perform interworking of a SIP-based public network with an H.323-based enterprise network, or any similar interworking between SIP and other VoIP implementations (even proprietary). In this case, signaling gateways are used to convert SIP signaling into H.323 or other signaling (or vice versa).

Signaling gateways may also be used to convert MTP/SS7–based transport of ISUP or BICC into an IETF SIGTRAN working group–defined transport of ISUP or BICC. SIGTRAN defines a protocol stack based on the Stream Control Transmission Protocol (SCTP [44]) and one of the adaptation layers, for instance, M3UA [45], which stands for MTP3 User Adaptation. This may be useful in trunking applications over IP networks, where both signaling and media need to be carried over IP.

When interworking between IPv4 and IPv6 is required, it can be implemented by means of a SIP Application Level Gateway (ALG) (similar to a way to support Network Address Translator [NAT] traversal, discussed later, in the section " Peering, Interworking, and Roaming"). This ALG function is commonly deployed at peering points also to implement some firewalling rules and to open "pinholes" in firewalls allowing media traffic to traverse the boundary between networks.

Address Resolution and User Identification   In order for SIP signaling to reach an IP endpoint where the desired SIP UA is located, it is necessary to know how to translate the identities carried in SIP messages (SIP user identity or Tel URI) into a valid destination IP address. As we have already mentioned, the format of SIP user identities is similar to the format of e-mail addresses, so a similar approach to address resolution based on DNS is adopted for SIP.

In addition, users may be subscribers of a PSTN service, and as such be given an E.164 number. These users may also be identified by a set of identities associated with other services (e.g., SIP-based VoIP, e-mail, etc.) they use in IP networks. In order to deliver advanced, converged, telecommunication services in this case, a telephone number must be mapped to a SIP URI.

The ENUM (RFC 3761 [46]) approach can be used to resolve a telephone number to an IP address or a list of IP addresses users can have their SIP calls or e-mails routed to (it could be different devices or different locations at the same time). This resolution is performed by DNS lookup of the Naming Authority Pointer (NAPTR) record in the domain "e164.arpa" of the telephone number associated to the user. More precisely, if a user has a telephone number +1-666-777-8888, the DNS query will look up 8.8.8.8.7.7.7.6.6.6.1.e164.arp. Here is an example of the NAPTR record returned in the DNS query (adapting an example provided in RFC 3761):

```
$ORIGIN 8.8.8.8.7.7.7.6.6.6.1.e164.arpa.
   NAPTR 10 100 "u" "E2U+sip" "!^.*$!sip:username@example.com!" .
   NAPTR 10 101 "u" "E2U+h323" "!^.*$!h323:username@example.com!" .
   NAPTR 10 102 "u" "E2U+msg" "!^.*$!mailto:username@example.com!" .
```

This example tells us that the domain 8.8.8.8.7.7.7.6.6.6.1.e164.arpa is preferably first contacted by SIP, second, by H.323 for voice, and third, by e-mail for messaging (as indicated by the order of values 100, 101, and 102 in the NAPTR records).

The final step in the resolution process is to find out what node to contact for each of the preferred contact options specified in the NAPTR record.

Clearly, address translation is the essential building block of convergence that lets both fixed (PSTN and VoIP) and mobile (circuit and packet) communications come together and makes users reachable at multiple contact points using a variety of communication technologies over a number of different access networks. All the users need is a telephone number that can be resolved to all the URIs specific to the different services. This is really one of the most powerful tools available to service providers to enable advanced converged communication services.

Voice over IP Deployment

Rolling out a VoIP service in a carrier's network is one of the most challenging migration and network transformation projects that the industry has faced. While the opportunities are undoubtedly significant, in terms of network operations, efficiency, and broadening of the potential service offerings, the challenges are also numerous. In this section we consider a number of possible solutions to common deployment issues, after having provided a high-level view of the rollout models that have been used by the industry in recent years.

VoIP Deployment Trends   The typical starting point from which incumbent service providers are deploying VoIP technologies and services is an existing PSTN infrastructure offering POTS and ISDN services, since only in very recent years have DSL-based broadband access services to the Internet boomed. Some operators are also offering innovative interactive services such as video on demand (or in some cases IP TV services) and converging services to provide *triple-play* solutions (including voice, video, and data) based on the bundling of traditional circuit voice services with TV broadcast and data access. The addition of cellular wireless service to such bundles leads the way to *quadruple-play* offerings.

Overall the following trends can be identified:

- Circuit-switched services need to continue to be supported. As the network migrates to VoIP for early adopters, the majority of the customers stick with analog phone service despite increasingly signing up for high-speed data services.

- It makes sense to initially provide VoIP by allowing customers to keep their existing analog phones and maintain the same user experience (with the exception of new features such as Web-based Voice Mail access and preferences configuration).

- Over time, new, more feature-rich equipment built specifically for VoIP and services utilizing newly available functionality may be introduced. Fixed communications will lead the way in migration from circuit to packet. The ensuing transition of cellular wireless communications from circuit-switched to packet VoIP telephony is also inevitable.

- When all services transition to IP, it makes sense to converge them into one IP-based network spanning across wireless and wireline domains to deliver customers a uniform access-independent experience. That will enable a fully integrated quadruple play using a single IP-based network infrastructure, an ultimate FMC goal.

VoIP Network Models   In the early days of VoIP service (the late 1990s), the ITU defined a "server model" for VoIP (also captured in Rel-4 of 3GPP specifications). This model is based on using BICC between media gateway controllers to interconnect media gateways (see Figure 3.6) over an IP- (or ATM-) based backbone. These media gateways act as the interface toward POTS or ISDN customers' access line termination infrastructure, toward the PSTN, and perhaps toward mobile networks supporting circuit-switched telephony.

Clearly this model, while viable, does not enable offering VoIP to residential customers but just supports the existing PSTN services over backbone networks based on ATM or IP. Although this approach (defined also as "3GPP Rel-4," due to the fact 3GPP standardized a server-based architecture for the CS core of UMTS in its Release 4) constitutes a way to use a common IP (or ATM) transport backbone to support current services, it does not allow the delivery of more advanced applications. So, it only enables the first of the four trends quoted above.

The limitation of this model was quickly recognized, prompting the industry to move on to the "next big thing" by using the developments of the 3GPP IMS effort. On a very high level, the IMS is a SIP-based multimedia services–capable system, which enables a service provider to offer *IMS-compatible* multimedia applications to IP endpoints supporting SIP UA capability. ETSI TISPAN and 3GPP have entered in a cooperation agreement to also define the use of IMS to deliver *traditional* (or legacy) wireline service to both users migrating to IP and those still attached to a traditional analog or ISDN telephone system, under what is called the NGN

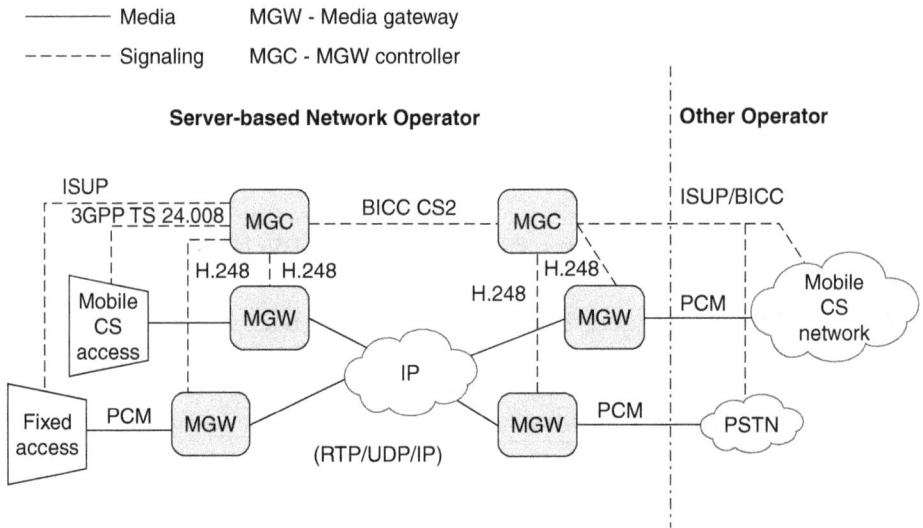

Figure 3.6   The ITU-T "server model"

(Next-Generation Network) program. At the time of writing, TISPAN has released NGN R1 (Release 1) and is in the process of defining NGN Release 2, while transferring the IMS-related parts of its work to 3GPP, in order to achieve the goal of a "Common IMS" shared by fixed and mobile systems.

TISPAN NGN provides the support for current POTS and ISDN customers, as well as customers using SIP-capable IP endpoints, using a common IMS core. Moreover, since the IMS core can also be used to serve wireless customers, this network model enables all of the four trends mentioned earlier and true convergence between fixed and mobile networks.

These NGN properties and its potential was quickly recognized by the industry, so the IMS, originally defined for 3G wireless, was quickly adopted by fixed service providers, both PSTN and cable multiservice operators (MSOs). This trend is further confirmed by the interest of CableLabs, a cable industry research and recommendations-setting body, and their involvement in 3GPP Rel-7 IMS specification activity. This activity and the increasing cooperation between different standards bodies provides clear evidence of the developing harmonization of the various branches of the telecommunications industry, not only happening at the infrastructure level but also in definition of the end-user services as exemplified by Figure 3.7.

VoIP Deployment Enablers   While the transition to VoIP promises to come with a way to reduce network operations costs and deliver new innovative services, a few deployment issues need to be sorted out, such as those related to Network Address Translator (NAT) and firewall traversal, delivery of adequate levels of QoS necessary for the different services, and security.

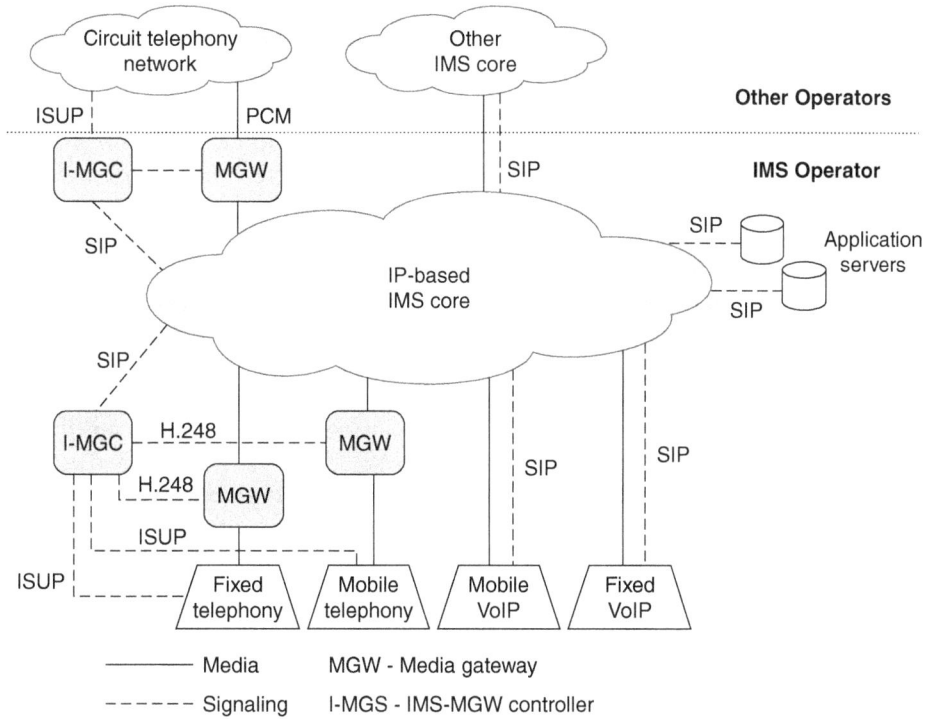

Figure 3.7   The IMS-based model

NAT Traversal   NAT is widely deployed in the Internet today to save on IP address space required to connect to the Internet. It is, for instance, commonly used in residential environments to connect multiple computers on the home network to the Internet via a DSL router. The DSL router supporting NAT assigns private addresses using DHCP to devices on the network and translates these "internal" addresses into a single IP address on the outside connection toward the Internet, multiplexing the "internal" addresses using port numbers.

The NAT traversal problem arises because the IP address and the port numbers used for VoIP signaling and media are transmitted within the SIP message, so the NAT cannot change them, as the devices supporting NAT (typically routers) operate at OSI Layers 3 and 4, and as such cannot manipulate the Application layer by default and cannot adapt the SIP header information following IP address and port number translation.

Therefore, whether a softphone, hardware VoIP phone, or analog phone connected to terminal adapter is used in the residential network, something needs to be done to overcome this limitation. Matters are complicated further because there is no such thing as a "standard NAT"; thus, many solutions may be required to address corner cases in proprietary NAT implementation strategies. Here we describe the most popular of these approaches.

STUN (standing for "Simple Traversal of UDP Through NATs," defined in IETF RFC 3489 [47] or "Simple Traversal Underneath Network Address Translators," as per the current revision of RFC 3489) is a standard designed to enable devices to figure out how they are seen from the Internet. A SIP UA device can use the information provided by a STUN server on the Internet (configured in the SIP UA similarly to a DNS server or a default gateway) in the SIP signaling it generates.

STUN, however, does not work well if the NAT opens different mappings for each new {IP address + port} combination ("symmetric NAT" implementation), so the information the STUN server provides may be valid only for the SIP UA-STUN server combination, which is clearly not suitable to initiate communication with another SIP UA or SIP proxy. So STUN can work well if the NAT device is STUN-compatible (no symmetric NAT) or if the UA is always using a SIP proxy as the first contact point of any SIP-based communication and the SIP proxy also acts as the STUN server.

A solution by which a client can obtain a transport address, which can be used to receive media from any peer, is to use a relay server that resides on the public Internet to relay data. Traversal Using Relay NAT (TURN) [49] method allows a client to obtain IP addresses and ports from such a relay server. Since this intermediate server is a costly solution to providers, it is normally not preferred over STUN. Interactive Connectivity Establishment (ICE) [48] can be then used to discover whether TURN is needed, and if not, STUN can be used instead.

A SIP-aware NAT, acting as an Application Level Gateway (ALG), may also be used as a solution for new deployments of NATs in new networks, but this solution can not be used for deployments where NATs are already embedded.

QoS   Another topic of extreme relevance in the deployment of VoIP is the support for guaranteed QoS to provide users with an adequate level of service. There are three main components (or requirements) in delivering QoS:

- Provisioning and engineering of resources for the different expected traffic classes
- Differentiation of traffic in the network based on traffic class
- Admission of users to use network resources provisioned to handle a specific traffic class

The first component is supported using classic network- and traffic-modeling techniques to predict how services need to be delivered. To be viable, the model supporting multiple traffic classes will need to be backed up by the provisioning of adequate network resources. Trend analysis techniques may also then be used to understand how well the model matches reality and to correct predictions once the service is deployed.

Different IP traffic classes may then be associated with different Differentiated Services Code Point (DSCP) values in the core. The concept of differentiated services (DiffServ) [39] is quite simple, as it allows core routers to differentiate traffic classes simply by looking up the Differentiated Services (DS) field in the IP header [50], to determine the per-hop behavior (PHB); that is the way to perform traffic handling for a specific traffic class at an IP router. As a result, Internet routers supporting DiffServ schedule traffic and manage queues by reading the value of the DSCP. Unlike RSVP-based signaling, this approach scales well in the core, so DiffServ is now adopted as the IETF standard to provide traffic differentiation in IP networks.

VoIP traffic is normally assigned a strict priority queue in core routers and given the DSCP field associated to the Expedite Forwarding (EF) PHB (that is, the PHB assigned to traffic that needs to meet real-time constraints).

Addressing the first two QoS requirements would not be sufficient if the third requirement were not met. Recently, TISPAN, ITU-T, and 3GPP developed methods to control QoS policy and charging rules to address the need for network resource admission. In TISPAN and ITU-T, the mechanisms defined to perform resource admission control have been named RACS (Resource Admission Control Subsystem) and RACF (Resource Admission Control Function), respectively. 3GPP in turn came up with the policy and charging control (PCC) framework (specified in 3GPP TS 23.203 [52]). All these standards approaches are currently subject of harmonization.

The operation of both RACS/F and PCC is based on controlling the use of the network resources at a gateway between the access network and the IMS core, via a dedicated server (called, in 3GPP, PCRF—Policy and Charging Rules Function—if it also handles charging rules or PDF—Policy Decision Function—if no charging-rules handling is entailed). Figure 3.3, shown earlier, conceptually represents the role of this admission control function.

Security   Since voice communications may deal with confidential matters, it is a requirement to support strong VoIP confidentiality protection. In IMS, IPsec (a standard defined by the IETF in [53] to provide ciphering and integrity protection of IP packets) is adopted to encrypt signaling between the SIP client and the network. IPsec with the Encapsulating Security Payload (ESP) header (defined in IETF RFC 4303 [54]) operating in IPsec transport mode has been selected for IMS in cellular environments. Note that IPsec transport mode protects just the content of the transport header and payload, not the packet IP header.

By definition, NATs do not allow traversal of IPsec packets, since the NAT device changes the IP address in the outer header of an IPsec-protected packet but cannot change the encrypted checksums of the IP addresses. In case NAT is applied in a wireline network between the user and the IMS core, UDP encapsulation (defined in IETF RFC 3948 [55]) of IPsec-protected packets is used to traverse NATs. It is also possible to apply encryption to the RTP stream directly via SRTP (Secure RTP, IETF RFC3711 [56]), a secure profile of the RTP protocol.

Another typical security requirement is the need to keep the network protected from external attacks while allowing for SIP-based VoIP to traverse firewalls. To achieve this, it is necessary to add to firewalls the SIP Application Level Gateway (ALG) capabilities. The ALG capability permits the opening of pinholes to let the media components traverse the firewall based on interpretation of the SIP signaling exchanged between the inside and the outside of the network that needs to be protected. It is often mentioned that the firewalls used in these cases should be "SIP-aware," in that they build state to permit media components to traverse the firewall, based on *awareness* of ongoing SIP sessions.

Since the NAT and firewall traversal, along with functions such as peering and inter-connection, are so important in service provider VoIP solutions, a market has emerged for products built to support QoS Service Level Agreement (SLA) enforcement, fire-walling, and lawful intercept in VoIP networks. These nodes, known as *session border controllers (SBCs),* also may act as SIP B2BUA and perform tight control on what ses-sions are allowed and what features can be supported by SIP endpoints (by breaking end-to-end transparency of SIP sessions).

While SBCs sometimes are a target of criticism by those adamant about the need to allow for SIP transparency end-to-end, others appreciate the level of control they allow service providers to exert on the usage of VoIP services in their networks. As always, there can be arguments for both camps, so we shall not take a firm position on this in this book, other than to state that we firmly believe SBCs (or equivalent SIP-aware entities, such as advanced firewalls implementations) are going to be present in many VoIP deployments for the foreseeable future to provide solutions to many challenges faced by service providers.

Peering, Interworking, and Roaming   The introduction of VoIP in many carriers' net-works inevitably comes with the need to define the way for these networks to exchange VoIP traffic. Since defining an ad hoc way to do that for each possible interconnect may result in excessive administrative and operational efforts for carriers, they normally come together at a national or regional level to define peering methods. Similarly, the exchange of traffic between PSTN and the VoIP network is defined by international standards (e.g., ITU-T Q.1912.5 [41] and 3GPP TS 29.163 [42]), which drive the de-velopment of devices such as media gateways and media gateway controllers, used in interworking applications.

In the UK, for example, the UK Network Interoperability Consultative Committee (NICC), including the service providers, both mobile and fixed, and vendors that operate within the UK, has defined the first regulated Inter-Service Provider VoIP interconnect in the world, under the auspices of OFCOM (the UK telecommunications regulating agency).

Indeed, the "Multi-Service Interconnect of UK Next Generation Networks" would act as a generic interconnect for PSTN/ISDN services that are on an NGN. The "purple release" of specifications from UK NICC is now available; it defines a generic inter-working and interconnect model between SIP-based and ISUP networks, where SIP networks are limited to run PSTN/ISDN simulation service (that is, SIP is used to implement ISUP-equivalent functions, based on the usage of ISUP over SIP [SIP-I], as defined as Mode C in the ITU-T document Q.1912.5 [41]).

The interconnection service at the Transport layer has been defined in document ND1611 [57] to be at Layer 2, following the tradition of network exchange points between service providers. This approach will use VLAN (virtual LAN) tags as multiplex-ing identifiers for the virtualization of connections between service providers over a shared infrastructure. Other documents specify the protocol-level details, e.g., for the transport of SIP signaling over TCP, UDP, and SCTP and the interworking between SIP-I and the UK ISUP. The logical view of this defined interconnect model is provided in Figure 3.8.

Operator A   Operator B

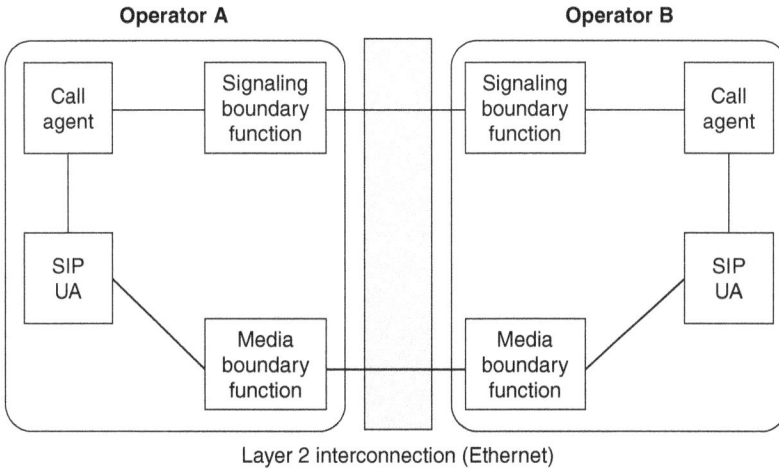

Layer 2 interconnection (Ethernet)

Figure 3.8   Logical view of the UK NICC–defined interconnection service

The SIP UA in Figure 3.8 can be a SIP-capable endpoint, such as an IP phone or a media gateway. The signaling border entities may be part of a session border controller device or any physical entity able to act as a SIP-aware firewall (the signaling border entity or the call agent must be able to open pinholes in the media boundary device, using unspecified interfaces as far as UK NICC is concerned, as the boundary entities may be implemented as a single device).

Another interesting aspect of VoIP is the intrinsic roaming capability that it can offer. IP networks allow, in principle, access to services provided by an operator with which the subscriber has a customer–service provider relationship (also known as a home network) from anywhere in the Internet. The VoIP service may therefore be used anywhere while the user is roaming, as long as the VoIP servers are routable from the point in the Internet where the user is connected (perhaps via the intermediation of a roaming partner of the home network). Service experience uniformity while roaming or while in the home network is assured by supporting service control in the home network, and by forcing SIP messages to traverse SIP servers located in the home network.

This model is used by the 3GPP IMS and is therefore adopted by all the NGN standards and network models based on IMS, such as TISPAN and CableLabs. The VoIP service provider may offer additional support for network access roaming if it also acts as an Internet service provider (ISP) and establishes roaming agreements with other ISPs. This may ensure adequate support in terms of QoS in the visited network and would provide an opportunity for visited network access operators to be included in the VoIP value chain.

Alternatively, the subscribers may opt for independent Internet access service, transparent to the VoIP home network ("bring your own broadband" or BYOB option). In that case the home network may need to offer services such as a STUN server for NAT traversal for its roaming customers. When the BYOB approach is used, the VoIP service provider must offer public "routability" to some boundary SIP proxies from any Internet endpoint (and not only from a set of well-known SIP proxies/servers belonging to roaming or interconnect partners) to ensure reachability.

Summary

In this chapter we have analyzed the state of fixed networking, paying special attention to the offering of both packet and circuit voice services such as those known as POTS, ISDN, and VoIP. While fixed networking is by no means limited to voice services, and indeed they no longer constitute a high-growth segment of telecommunications, these services are essential to the topic of FMC (where voice service continues to be the focus of the various converged offerings).

In the following chapter we are going to tackle the M in the FMC: mobility. We are going to discuss the state of today's wireless communications and address different technologies composing the space, including cellular, Wi-Fi, WiMax, and others.

4

The *M* in FMC

Give me a place to stand, and I will move the earth.

—Archimedes (287–212 B.C.)

Without a good understanding of fixed networks, which you already should have acquired by reading the preceding chapter, it would be quite difficult to fully appreciate the aspects of mobility we are going to delve into in this chapter. Indeed, much of the fundamental functionality and many architectural aspects of fixed networks have been borrowed and extended to create the core of modern mobile networks.

For instance, termination of a voice call in a mobile environment requires an interaction with a database storing subscriber profiles, the Home Location Register (HLR), much like the interaction that occurs in fixed networks with an SCP in order to route a toll-free phone call to the correct call center's call distributors. This example well illustrates how lessons learned and concepts invented for fixed environments are reapplied in a mobile setting, albeit with the necessary adaptations.

In this chapter we will identify these similarities and differences, and explore the most common cellular and noncellular radio access systems that may commonly be part of FMC solutions. The knowledge acquired in this and the preceding chapter will then make it possible for you to fully appreciate the ensuing discussion on convergence.

In the end, *fixed* and *mobile* communications can be seen (and they are) as two different facets of the same technology, which indeed have never been totally independent, so convergence for them is a consequential natural course of evolution. Thus, sooner or later, fixed and mobile networks will come together—the process already evident in the industry with the launch of services allowing use of a mobile phone at fixed-line rates, and using fixed-line access, when in the home—with FMC being one of the technologies hastening this "homecoming." To paraphrase Archimedes' famous statement: "Give me a fixed network and I can build mobile networks around that."

Mobility

While mobility is not only about wireless or radio access systems (as we have established in earlier chapters, there is quite a good distinction between *wireless* and *mobile,* wireless referring to the type of access and mobile to the type of service), wireless systems are usually at the core of mobile communications, so in this chapter most of the discussion in fact revolves around such solutions. It should be noted that many wireless access systems such as Wi-Fi do not natively support mobility. In fact, most of today's applications have relied on Wi-Fi mainly for "cord cutting" while accessing fixed networks; that is, this has held true until the recently introduced uses of Wi-Fi for metro data coverage and wireless VoIP broadened its scope, albeit without the support for macromobility allowed by cellular systems.

In the future it is also likely that Wi-Fi and other technologies such as WiMAX will be used to connect to mobile packet cores (à la 3GPP TS 23.234 [82]), which is not surprising given their broadening support on cellular phones. Before turning to Wi-Fi, however, in the sections that follow we will first address cellular systems and then introduce WiMAX access, paying special attention to its mobile version.

Mobile Communications Systems Fundamentals

One of the fundamental characteristics of mobile systems is that, like fixed systems, they include a core network, the elements of which act as an anchor for mobility. The core network also contains points of interworking with other (fixed and mobile) networks and connects to an access network needed to support the radio interface technology that characterizes a specific mobile system.

Core and Access

Note that different mobile systems, defined by various radio interfaces and access networks, may still share the same core network. For instance, it is possible to use the same core network to support the Global System for Mobile Communications (GSM—originally an acronym for *Groupe Spécial Mobile* and now the most widespread cellular technology) and the Universal Mobile Telecommunications System (UMTS).

Core Network Functions   The core network normally includes the "intelligence" necessary to make the system operate. The core provides the admission control to wireless services via access authentication and stores subscriber profiles, letting the core determine the specific service set and treatment to be applied to a user. These are examples of the questions the core can decide on:

- Is the user allowed to roam to another network?
- Are there areas where the user cannot receive service?
- Is the user in and out of a "home zone"?
- What Quality of Service (QoS) for packet data services is the user is entitled to?
- What calls is the user entitled to receive or place, and which are barred?

The core also provides *mobility management* of the mobile terminal in idle mode, that is, at times when it is not engaged in a voice call or in the active transmission and reception of data. The mobility management function allows tracking the user's whereabouts, by means of location updates the terminal issues periodically or when it crosses boundaries of "location areas" as the user moves.

Further, the core participates in call control for speech services and session control for multimedia services as well as providing infrastructure that is ancillary to the delivery of services (such as location information). The core network is also the main source of charging information.[1] Finally, mandated services like lawful interception rely on the core network to provide support for tracking the location and specific behavior of a user, log its activity types, and deliver the content of communications to the authorized agencies.

All this intelligent functionality (typically supported by different servers and network elements) can be quite computationally intensive, and concentrating these in centralized locations provides efficiency gains in large part for statistical reasons (not all users will need to use the same resources at the same time, so centralizing some functions will increase efficiency). Another reason for centralization is the need to provide a "point of access" for external networks to the mobile system services. Examples of this include scalable call termination and roaming interfaces (on roaming interfaces, it is in fact necessary to minimize the number of peering relationships to simplify roaming operations).

Access Network Functions   Typical functions of the radio access network are Physical-layer termination and Link-layer termination of radio interface protocols, ciphering and header compression,[2] mobility management within the radio access, paging, broadcast information distribution, and scheduling.

---

[1] In some systems, like in UMTS, the RAN cooperates with the core by delivering unsent data volume reports, but this feature is most likely to be discontinued in future systems, as it either was not implemented in most cases by vendors or was not used by providers, and of course the increase of data rates will make the value per bit unsent smaller, so it will be less important to count unsent data.

[2] In some technologies like GSM, ciphering and header compression for data sessions are done in the core network.

It should also be understood that for a mobile system to provide a satisfactory user experience, it must offer extensive coverage over large geographical areas. There is also the need to keep the transmission power of terminals and the complexity of receivers low to allow for longer battery life and lower terminal cost. To assist in that goal, the reach of the base transceiver stations (BTSs, which include the antennas in the network, radiating signals to mobiles and receiving signals from them) has to be limited (also to allow for frequency reuse, as explained later in the chapter), so the number of elements in a typical radio access network is much higher than in the core, and this is where most of the capital investment is needed.

Since BTSs represent the largest number of elements in a mobile radio access network, the functions in the BTS tend to be kept to the bare minimum necessary to provide radio access. However, with the introduction of high-speed wireless data services, there is a trend to offload some of the functions that were suitably kept in the access nodes higher up in the radio access network hierarchy down to the BTS. There is a host of technical reasons behind this, mostly linked to the delay introduced by the traversal of a backhaul network to a controller, which would make, for instance, running some retransmission algorithms less effective.

This trend of assigning additional functions in the BTS to handle data access has been taken to the extreme by the introduction of the concept of "Home BTS" or *femtocell,* which is promising the delivery of "personalized" wireless coverage in residential or business premises not reachable by the existing WAN infrastructure of BTSs. This technology is driving the use of IP connectivity to the BTS in order to reuse the infrastructure already available in those premises, and the simplification of the protocol layers supported on the interface between the core and the *femtocells,* to minimize the cost of backhaul.

Circuit and Packet

Mobile networks, like fixed networks, can provide a variety of services, including circuit-switched (CS) and packet-switched (PS) services.

Circuit services include speech, video, and circuit-switched data (akin to landline modem dial-up—fast becoming a thing of the past with the rise of broadband—based on dialing a circuit connection to a remote access server (RAS) and running PPP over it).

Packet services include high-speed data used for access to Internet and private networks, portal access, Web browsing, instant messaging and presence, and other innovative services, including IMS, multicast, and broadcast.

Circuit-switched services are supported by all traditional *cellular* systems, whether digital or analog (the latter actually have been discontinued globally, with possibly a few rare exceptions). However, not every *mobile* communication system supports circuit services. For instance, the evolution of the 3GPP system foresees that all services will be packet based and, more specifically, IP based. Other mobile systems such as WiMAX and CDMA EV-DO revA were built as packet only from the ground up.

On the other hand, packet-switched services were not supported by analog systems or by initial releases of digital systems, as at the times these systems were introduced, the demand for data services was not sufficient to justify the additional effort to define packet data support in cellular systems. This functionality was added in subsequent system releases as an evolution of both the radio and the core network of these systems.

The need to support both packet- and circuit-switched services requires the core and access of a mobile-system to support features that are often very different in nature. The mobile-system features dedicated to CS services are often identified as the "CS domain," whereas the set of features devoted to the support of PS services are defined as the "PS domain."

Having thus explained the fundamental differences between PS and CS, we are now ready to proceed with an overview of the specific mobile systems, starting with the discussion of the traditional cellular infrastructure.

Cellular Systems: The Sky Is the Limit

The concept of *cellular* has been invented to cope with one fundamental issue in wireless transmission technologies operating in a licensed spectrum.[3] The issue is the scarcity (and cost!) of the radio frequency (RF) bandwidth. When an operator is allowed to provide wireless service, a regulatory authority assigns it portions of the RF spectrum (known as "bands") to use for radiating and receiving signals from its subscribers.

Sometimes operators engage in fierce bidding to buy some of this spectrum, which results in staggering amounts of money being transferred to the government (or, in some cases, to previous owners of the spectrum bands). Therefore, the need to maximize the efficiency when using this precious resource, as well as the need to overcome basic physical limitations of wireless transmission, has led to the invention of what is called *frequency reuse*. With frequency reuse, multiple transmitters can use *the same* radio frequency to send and receive data as long as they are "sufficiently remote" so as to avoid interference. That is, a given (RF) channel can be used over and over again and thus cover an extremely wide area and support a very high number of simultaneous transmissions.

The concept of "sufficiently remote" requires some degree of clarification here. It is a common experience for everyone that the coverage of a radio station can be quite broad even without repeaters. The reason for this is that the radio station transmits at high power levels. In cellular systems, by contrast, the goal is to control the power of transmission so that interference is avoided between areas where the same frequency is reused. So, the concept of "sufficiently remote" is relative to the possible reach of a signal transmitted at a given power level. One of the most important characteristics of frequency reuse is the *reuse distance*. Intercell interference can be limited by changing

---

[3] Licensed spectrum is spectrum for which an operator needs to obtain authorization from a regulatory authority for the use of the radio frequency (RF) it needs for the operation of the wireless service.

the reuse distance and the power levels of BTS transmission. The resulting combination of power control and frequency planning is used to fine-tune the cellular network in a given area.

The elementary area of coverage in a cellular system roughly centered around the BTS is called a *cell*. Each BTS can use a subset of the RF channels the operator is allowed to operate. These channels cannot be reused in any neighboring and potentially interfering cells (that is, any cell within the "reuse distance"). With the introduction of cellular systems based on Code Division Multiple Access (CDMA) technology, there have been ways to advance the use of digital technology to reduce interference while allowing for greater reuse levels and thus increased capacity. Also, other techniques such as cell "sectorization" (using directional antennas, thus adding space division to frequency division, in order to maximize frequency reuse) and dynamic channel assignment to cells (thus introducing the concept of time division in addition to frequency division) have made it possible to further increase capacity by allowing higher reuse factors and availability of more channels per BTS.

In summary, the cellular systems have made it commercially viable to support mobility of a large number of subscribers in a wide area at a reasonable cost practically without any limitations as long as a sufficient number of base stations is installed. It is no surprise that, given the relative ease and speed of deployment allowed by cellular systems, and the reach of customers in even the remotest areas at reasonable costs, cellular communication is now a dominant and growing sector of the telecommunications industry. This sector is, however, relatively nascent in comparison with fixed networking (it is only a few decades old, compared to the more than one-hundred-year history of public wireline telephony), but it has dramatically changed the way we communicate. Its evolution has also been quite rapid from the early days of analog cellular telephony to the recent days of high-speed wireless data. The next section summarizes this evolution.

Cellular Generations

The evolution of cellular communication systems has been commonly identified via a series of generations (1G, 2G, 3G, and, by implication, 4G and yet further generations). This evolution has been somewhat linked to the evolution of computing. When digital computing was expensive, it was considered unfeasible to digitally process speech and use digital transmission of digitally encoded voice over the radio interface, in large part due to the high handset power requirements and limited memory capabilities. Therefore, the natural choice was to keep wireless telephony entirely in the analog domain.

1G Analog cellular systems are commonly known as 1G (or first-generation) cellular systems. 1G was introduced in the late 1970s and early 1980s. These systems delivered almost exclusively speech services; albeit some low-bit-rate data services were available as dial-up connections via wireless modems.

Analog systems are almost everywhere being phased out, and those that are not will probably be retired very soon. From a practical standpoint, the investment in these systems has stopped on both vendors' and operators' sides, and for this reason they are not considered in the scope of convergence, which mostly lies in an area where the industry is going to develop.

Typical 1G systems include:

- **AMPS**   Advanced Mobile Phone Service, adopted mainly in the America, using FDMA transmission in the 800 MHz band.

- **TACS**   Total Access Communication System, adopted mostly in Europe. TACS was similar to the AMPS system. There were various flavors of it; for instance, in the UK, ETACS (Extended TACS) operated in the 871–904/916–949 MHz band, and Narrowband TACS (NTACS) operated in the 860–870/915–925 MHz band (by using narrower channel spacing, it supported more channels for the same amount of spectrum).

- **NMT**   Nordic Mobile Telephone, deployed in many European countries, was first launched in the Scandinavian region (as its name may suggest) in 1979, and it was the first analog cellular system operating in both the 450 MHz and 900 MHz bands. The operation in the 450 MHz band affords particularly good coverage due to the propagation properties of radio waves, so it has been kept alive for quite a long time in Scandinavia, often used by boat owners in the Baltic region, or rural area dwellers, where it was not economical to deploy systems in different bands. The "450" frequencies are therefore particularly appealing for these reasons, and digital systems offering the capability to cover these subscribers (GSM, CDMA, and also 3G) have been offered to refarm this system.

1G systems were deployed on a country-by-country basis and did not offer international roaming. This became an apparent issue, especially in Europe where there were two different analog systems operating in different frequencies and according to different rules. This clearly was creating a problem for the development of the cellular industry, and therefore even at the political level it was clearly understood that the development of competing digital systems in that region would have been a critical mistake.

Beyond forcing compatibility issues and hampering roaming capabilities for European citizens, this would have fragmented the market and made it more difficult for European vendors to compete globally. European operators would also be unable to benefit from effects of scale for both equipment and, most important, handsets. In summary, this was at the core of the reason why European regulatory bodies encouraged operators and vendors to come together and agree on the common digital system, which was later named GSM.

2G   Cellular digital systems in the first wave to appear in the mobile communications industry are known as 2G systems. These were rolled out worldwide in the course of the 1990s. These systems are characterized by the digital transmission of all services, including voice, which is digitally encoded for transmission over the radio interface.

Initially they were conceived to deliver circuit services only, but then their evolution to deliver packet data services was devised (these systems' evolution to support packet data later became known as 2.5G).

Typical examples of 2G systems include:

- **GSM** The Global System for Mobile Communications initially is the European 2G system, specified by ETSI and currently maintained by 3GPP. It is now adopted in practically every country or region of the world, except Japan, Korea, and a few others. GSM was initially operating in the 900 MHz band only, but then operation at 1800 MHz, 1900 MHz, 450 MHz, and other frequencies was also defined. GSM uses a time division multiple access (TDMA) multiplexing over-the-air interface. Its 2.5G system evolution is known as General Packet Radio Service (GPRS), and its 2.5G air interface evolution is known as Enhanced Data Rates for GSM Evolution (EDGE). EDGE has also been accepted as an IMT-2000 technology, so technically it is also part of the "3G" mobile systems; albeit it is deeply linked to its 2G heritage.

- **TDMA** Time division multiple access, defined in TIA IS-136 [96], or D-AMPS (for Digital-AMPS), has been adopted in North and South America. It uses a time division multiple access technology (like GSM), over-the-radio interface, but unlike in GSM, this technology was not globally adopted, so it has been largely phased out in favor of GSM in every region where it was deployed (except some South American countries, where it may still be available for a while).

- **PDC** Personal Digital Cellular (PDC) is the Japanese standard for digital cellular telephony. In Japan, this is quickly being replaced by 3G systems, and it has received quite phenomenal competition from a Japanese cordless telephony standard (PHS—Personal Handy-phone System), which has been made available in major urban areas to the large population of pedestrians and users of public transportation systems. PDC is also a TDMA system, replacing the analog NTT and JTACS Japanese systems. It operates in the 800 and 1400 MHz frequency bands.

- **CDMA** Code Division Multiple Access, specified in standard TIA IS-95A [98], uses a technique to assign to all transmitters their respective, specific, bit vectors, which are mutually orthogonal so that a receiver can recover the signal transmitted even if multiple transmitters use the same carrier at the same time to modulate the RF signal. This standard is now maintained by 3GPP2 and has been adopted in many countries in Asia and South America, after its initial deployment in North America. The evolution of IS-95 to support packet services is known as IS-95B, and it supports data rates up to 64 Kbps.

3G Similarly to what happened in Europe during the migration from 1G to 2G, during the migration from 2G to 3G it appeared that having many flavors of 3G standards across the globe was not advantageous. So various standards organizations from different countries came together to form the Third-Generation Partnership Project (3GPP) to define a standard based on the evolution of the GSM core network and W-CDMA radio transmission technology (the most promising and efficient at the time).

In North America, this evolutionary path to 3G was not received too well, though, as major CDMA operators (and their main suppliers) perceived it not to be protecting their investment sufficiently. So a competing partnership was formed, called 3GPP2, developing a 3G standard based on the evolution of IS-95 CDMA radio transmission technology and the ANSI-41-based core network.

With the creation of 3GPP and 3GPP2, the intent of achieving a global cellular standard for 3G has failed. However, there was a considerable momentum toward harmonization for both competing standards as the number of different systems went down from four in 2G to two in 3G.

The 3G standardization activity took place under the supervision of the IMT-2000 in ITU-T, which was assigning frequencies and acknowledging technologies suitable for the 3G standards (also known hence as IMT-2000 technologies). 3G Systems, launched after the year 2000, promised to offer faster access to the Internet and other data services with typical speeds ranging in the hundreds of Kbps. They also offered circuit services like speech and real-time video. The systems resulting from the efforts of the 3GPPs were:

- **UMTS** The Universal Mobile Telecommunications System, which has been specified by 3GPP, is a multiradio interface system, which includes satellite communications, but its most widespread use for now is based on a W-CDMA radio interface, and it operates in different frequency bands, depending on the region of deployment. W-CDMA comes in FDD (frequency-division duplexing) and TDD (time-division duplexing) flavors. In FDD, uplink and downlink signals are allocated to different frequencies, while in TDD they share the same frequency on a time-division basis. UMTS offers both packet data and circuit services. In China, the TD-SCDMA (time division–synchronous code division multiple access) radio interface has been selected to constitute the homegrown UMTS radio interface. Japan launched 3G UMTS before any other country, due to the leadership of DoCoMo in 3G UMTS-based standards with its Freedom of Mobile Access (FOMA) system.

- **CDMA 2000** CDMA 2000 evolves IS-95 to include additional service capabilities based on packet data. It is a direct competitor of the other major 3G standard, UMTS, and operates at 450 MHz (used to refarm NMT, as mentioned earlier, along with GSM operating in the same band, and in developing countries to allow better coverage), 850 MHz, 900 MHz, 1800 MHz, 1900 MHz, and 2100 MHz. CDMA2000 1XRTT has been the first step in the evolution to 3G, which has been recognized as an IMT-2000 technology, and also the first 3G technology to be deployed worldwide.

Since the first releases of the 3G systems, they have evolved quite significantly. CDMA2000 EV-DO (Evolution–Data Only or –Data Optimized, also known as the High-Rate Packet Data air interface) and EV-DO revision A (DOrA), HSDPA (High-Speed Downlink Packet Data Access) are examples of the 3G high-speed data services–capable systems that grew out of the original UMTS and CDMA2000 standards.

If the industry continues to follow the trend established during the transition from 1G to 3G in the transition from 3G to 4G, we should expect to see only one global cellular system based on a common core and radio transmission technology. While this is not yet a certain fact, all evidence suggests that this is going to be the case, as major operators of 3GPP and 3GPP2 networks seem to be converging on a single evolutionary path to 4G.

We must mention, however, another trend developing in parallel with cellular standardization. It appears that systems such as WiMAX (and, to a certain extent and with some additional limitation, Wi-Fi) are being positioned as yet other candidates for deployment in many operators' networks as a complement to or even a complete replacement for 3GPP and 3GPP2 cellular systems.

So, at a time when major cellular families seem to be on an eventual path of convergence, there is a potential for competing technologies (if only on the radio interface side) to be adopted by both the cellular and wireline operators. As a consequence, this may result once again in a plurality of systems. On the other hand, due also to the potential existence of such a plurality of access technologies that need to provide access to the same set of services, it is also recognized that using a single access-independent core is indeed valuable. As such, the industry is experiencing more and more a drive toward using a common core for all mobile (and nonmobile) systems, providing access to the same set of services, maybe based on IMS, thus validating the overall direction toward convergence.

In the following sections we provide a more detailed review of the mobile systems, which are best positioned to participate in FMC solutions. In this discussion we are going to focus on the relevant aspects of these systems such as core network details and operation, and characteristics of the access network that need to be considered when being converged with the fixed network in the context of practical FMC solutions.

3GPP Systems

Both the GSM 2G mobile system and the UMTS 3G system are specified and maintained by 3GPP. The GSM system was originally specified by ETSI under the drive of the European countries[4] to share a common digital system to enable easier intraregion subscriber roaming, but when 3GPP started its work to define a 3G system evolved from the GSM core, it became clear that it would have made perfect sense to also maintain GSM specifications and define their evolution within 3GPP.

Today it is possible to roam using GSM in more than 210 countries. GSM operates in the 900 MHz and 1.8 GHz bands in Europe and the 1.9 GHz and 850 MHz bands in the U.S. The 850 MHz band is also used for GSM and 3GSM in Australia, Canada, and many South American countries. UMTS was originally defined to operate in the

---

[4] The European Conference of Postal and Telecommunications Administrations (CEPT) created the *Groupe Spécial Mobile (GSM)* in 1982 aiming at developing a standard for a mobile telephone system that could be used across Europe.

1885–2025 MHz band for uplink and 2110–2200 MHz for downlink, though, in addition to these spectrum ranges, today it is commonly run on 850 MHz and 1900 MHz in some countries, notably in the U.S. by AT&T.

3GPP subscribers are identified by means of an International Mobile Subscriber Identity (IMSI), and they are assigned one (or more) MSISDN (Mobile Station ISDN) number, that is, an E.164-compliant telephone number. These are stored on a Smart Card known as the Subscriber Identity Module (SIM) card for GSM, or Universal SIM (USIM) card for UMTS. The smart cards are given to subscribers when they sign up for service, and they can be installed on any 3GPP-compatible handset for its activation with the service. The SIMs are not commonly used in 3GPP2 systems, where subscriber identity is linked to a specific terminal; albeit in recent years the capability to use a smart card has been added to CDMA standards and is now being deployed (especially in China, among other countries) with the smart card–based Removable User Identity (RUID).

The GSM and the UMTS systems share a common core network for both the CS domain (also known as circuit core) and the PS domain (also known as GPRS core, or packet core).

The 3GPP CS core   The 3GPP circuit core (shown in Figure 4.1) is based around the Mobile Switching Center (MSC), which acts as a switch for voice and circuit data calls and handles mobility management of "CS-attached" users. CS-attached users are authenticated and accepted by the circuit-switched mobile core and can therefore be handled by the 3GPP system. Authentication of 3GPP users is based on data and algorithms stored in the SIM. On the network side, the authentication function is based on data downloaded from the Home Location Register (HLR), specifically from the

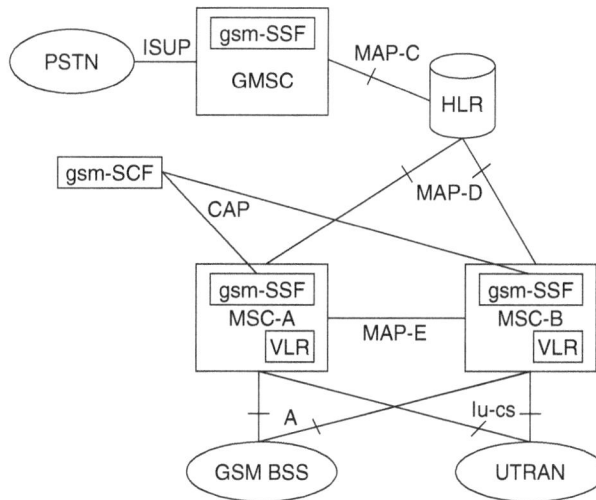

Figure 4.1   The 3GPP circuit core

authentication center (AUC) component of the HLR, into the serving MSC, and more precisely in the Visitor Location Register (VLR), also commonly a component of the MSC. The HLR and VLR are subscriber databases where the subscriptions to services, their parameters, and their activation status information are stored, along with the authentication information.

The HLR is always located in the home network, the network with which the subscriber has a customer-provider relationship. The VLR is located in the visited network. Since both the roamers' and home subscribers' data are stored in the VLR when a subscriber is registered with an MSC, the VLR effectively assumes a role as a cache of subscriber data, used to avoid continuous interrogations of the HLR, which would cause significant signaling load. Note that the VLR was defined to be potentially stand-alone, but in all practical cases today it is implemented within the MSC platform.

MSCs also host Customized Applications for Mobile Network Enhanced Logic (CAMEL) Intelligent Network triggers via the GSM Service Switching Function (gsm-SSF), which are armed to provide Intelligent Network–supported capabilities such as toll-free calling, caller ID, location determination, call forwarding, and so on, by interacting with the GSM Service Control Function (gsm-SCF).

In addition to these functions, the MSC often acts as a gateway to the PSTN, and as such it is identified as gateway MSC (GMSC). In this role the GMSC, as shown in Figure 4.2, would receive incoming ISUP call establishment signaling messages from the PSTN and query the HLR for the whereabouts of the user (more precisely, find out the MSC where the user is known to have last registered with the HLR).

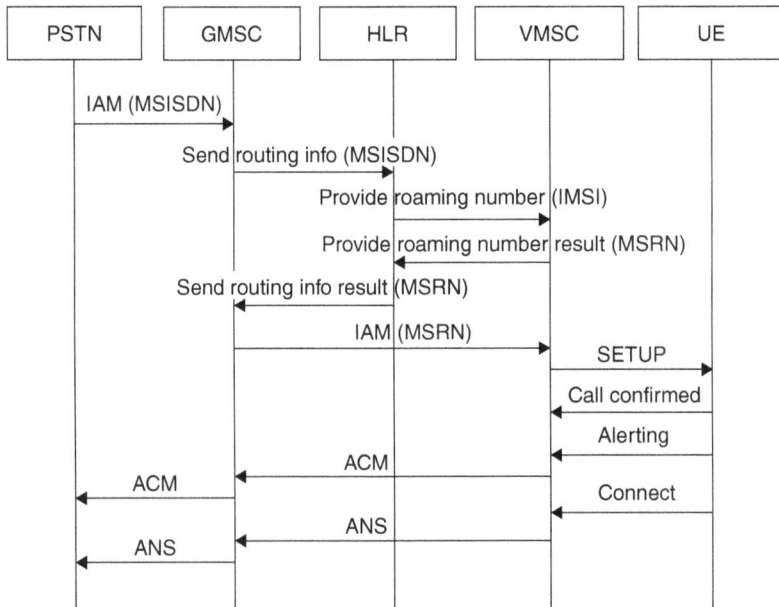

Figure 4.2 Mobile-terminated call setup

The HLR, as shown in Figure 4.2, then queries the visited MSC's VLR to obtain a Mobile Station Roaming Number (MSRN) for the subscriber; the VLR allocates one MSRN and returns it to the HLR, which passes it back to the GMSC. The GMSC would then use ISUP to establish a circuit using the MSRN as the destination number. When the call is set up, the MSRN can be released and used for another subscriber, thus limiting the number of E.164 addresses needed for mobile-terminated call routing at an MSC.

The approaches similar to the one used for such termination of calls have been used for a mechanism to route calls to the IMS or the CS domain in Voice Call Continuity (VCC—described in detail in the next chapter) for dual-mode terminals capable of receiving a voice call in the IMS-based PS or in the CS domain, depending on which network they are attached to. VCC enables the termination of calls in either domain and also enables continuing the call from one domain to another as the terminal changes the access technology it is camped on. VCC is therefore a potentially fundamental component in many FMC applications described in this book.

An MSC can act as a point of interconnection toward the PSTN for outgoing calls, so that they are optimally routed to the PSTN destination if necessary. MSCs interact with the HLR using the IS-41 [101] interface in CDMA and the Mobile Application Part (MAP [102]) interface in GSM. MAP interface variations include:

- MAP-D between VLR and HLR
- MAP-C between the GMSC and the HLR
- MAP-E between MSCs to prepare handover

The MSC also implements circuit-switched supplementary services based on settings and an activation status provided by the HLR (or retrieved from the VLR in the last visited MSC) when a subscriber is accepted by an MSC. This is in contrast to relying on a centralized execution of service logic as in the case of CAMEL-based services, which follow an Intelligent Network model of decoupling service execution from switching.

When integrating a CS network with a PS (for example, an IMS-based PS) network in a converged environment, the operator needs to make sure that services invoked and triggered in the CS network have their state updated or kept in sync in the PS network. Ideally, the services themselves should be evolved to be executed in a single centralized location (logical and/or physical). The example of such centralization is presented by IN and the use of IMS application servers, when dual-mode CS/PS terminals are being used.

Mobility Management   MSCs are directly involved in *mobility management (MM)* of users in the idle and active states of both GSM and UMTS, by handling location update procedures and by being involved in handovers between MSCs or between different nodes in the radio access directly attached to the MSC. When the inter-MSC mobility events occur, the new MSC updates the location of the subscriber with the HLR, so that the HLR knows how to route mobile-terminated calls for the subscriber, or how to enforce some asynchronous actions (e.g., purge the subscriber identity and profile from the last known MSC).

Handling mobility management implies interaction with the terminal, which is also necessary for call control. The GSM and UMTS systems use an ISDN-access-signaling-like interaction with the terminal for call control, specified in 3G TS 24.008 [180].

The interfaces to the radio access network of GSM and UMTS, respectively, as shown in Figure 4.1 earlier in the chapter, are called the A interface and the $I_{u\text{-}cs}$ interface. The $I_{u\text{-}cs}$ interface is the interface between the UMTS access and the 3GPP CS core, and it also has an $I_{u\text{-}ps}$ component for the PS domain core. The A interface assumes an E1 (or T1) transport to be available, while the $I_{u\text{-}cs}$ interface assumes an ATM or IP-based transport.

Transcoding   The transcoding between the PCM encoding used in the PSTN and the voice codecs used in GSM (AMR, or adaptive multirate voice coding) happens in the base station controller (BSC). Specifically, the *transcoding unit (TRAU)* is a function of the BSC dealing with transcoding. The *A* interface therefore carries the PCM-encoded voice-over-TDM multiplexed channel.

In UMTS, this model changes, with transcoding taking place in the core or preferably avoided altogether with the Transcoder-Free Operation (TrFO) option, where the same encoding of voice is used between terminals in mobile-to-mobile calls. In the TrFO model, the voice quality is therefore quite substantially improved because it avoids several transcodings necessary in the end-to-end data path. For communication with the PSTN, the transcoding between the AMR [63] (or wideband AMR [64]) codec used in UMTS and the PCM codec used in the PSTN takes place as close as possible to the point of interworking with the PSTN.

Bearer-Independent Circuit-Switched Network (BICSN)   The UMTS TrFO feature has been defined in a server-based architecture introduced in 3GPP Release 4. By enabling packet-based transport of voice, it is possible to eliminate the need of transit switches in the core network and as such the need to convert AMR- or WB-AMR-encoded voice into PCM for switching in a classic TDM framing. This server-based architecture for the CS core, also known as a bearer-independent circuit-switched network (BICSN), is represented in Figure 4.3.

The attentive reader has already noticed that the difference between the classic architecture and the BICSN is that the MSC is split into a media gateway and MSC server components. MSC servers control the media gateway via the H.248-based M_c interface. The N_b interface supports the transport of the media components between gateways, and the N_c interface, based on BICC signaling, is used to let MSC servers interact with one another to establish calls or set up inter-MSC legs during handover. In addition, starting with Release 4 of the 3GPP specifications, the HLR has been replaced by the new functional element called the Home Subscriber Server (HSS) as a part of the IMS framework. The HSS extends the HLR functionality to support additional interfaces for the interaction with IMS entities, via DIAMETER-based interfaces.

The 3GPP PS Core   The PS core of GSM and UMTS, shown in Figure 4.4, is in many aspects similar to the CS core, in that there is a gateway entity, called the gateway

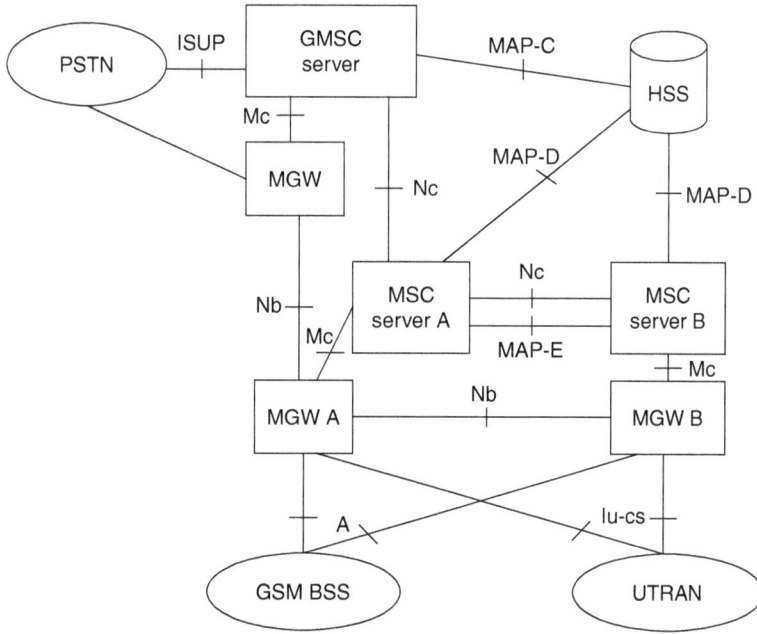

Figure 4.3   Bearer-independent CS network

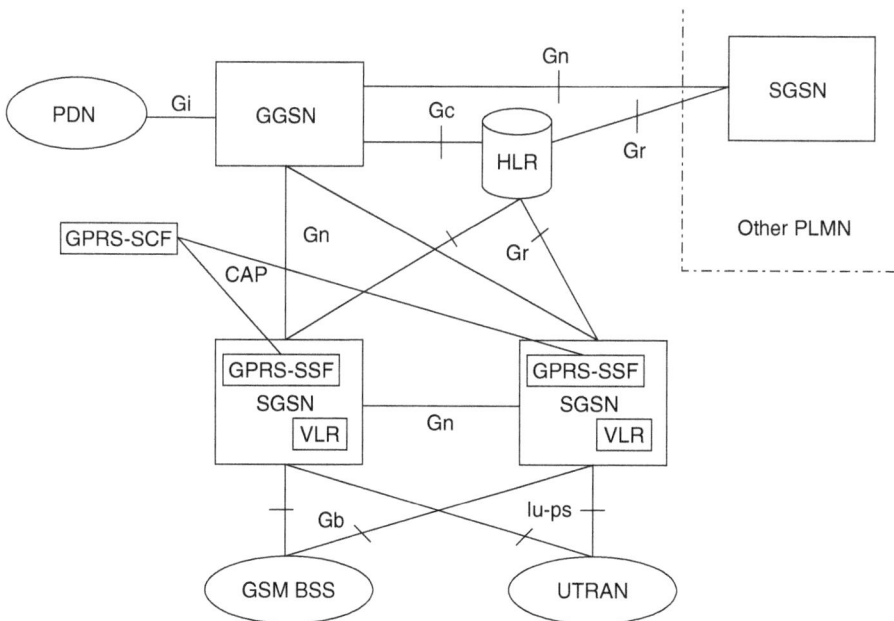

Figure 4.4   The GPRS core network

GPRS support node (GGSN), and a serving entity called the Serving GPRS support node (SGSN) supporting packet data mobility. GPRS itself stands for General Packet Radio Service, which is an extension of the GSM system designed to support packet data services. GPRS is optional to GSM but is an integral component of UMTS from the outset, so technically speaking, the support of GPRS core and services is "not an option," if we can say so, when deploying a UMTS system.

Similarly to the CS core, it is also possible to use CAMEL Intelligent Network services in GPRS, mainly to support prepaid charging models. This is based on the interaction of the SGSN-hosted gprs-SSF, where the CAMEL triggers are armed, with the gprs-SCF. Since the Intelligent Networking subsystem had its roots in the legacy circuit domain, it often does not provide optimal solutions for packet data applications. For example, since deep packet inspection and per-flow charging and QoS policies are better supported at the GGSN, the prepaid and other charging models and services traditionally associated with Intelligent Networking are now transitioning to being handled by the DIAMETER-based interfaces between the GGSN and servers specialized for these functions. This trend is also in alignment with other sectors of the industry (e.g., the RACS/RACF subsystems use DIAMETER-based interfaces for QoS policy control).

The interconnection between the elements of the GPRS core and the HLR (or HSS, in 3GPP Rel-4–based systems) happens via interfaces called G_r (between SGSN and HLR) and G_c (between GGSN and HLR). The G_r interface is used for user data download in the SGSN and location update, while the role of the G_c interface is linked to the support of the network-initiated data sessions feature. This feature works only for statically assigned IP addresses, and it has not proven to be very popular, as IPv4 addresses are a scarce resource and therefore are rarely statically assigned to mobiles. On the other hand, although IPv6 is supported by GPRS standards, and it does not suffer from the issue of scarcity of IP addresses like IPv4, this IP version has not been widely adopted in commercial deployments yet.

The interface between the SGSN and the UMTS access is called $I_{u\text{-}ps}$, and the interface toward the GSM access is called G_b. The $I_{u\text{-}ps}$ interface assumes that either IP transport or Asynchronous Transfer Mode (ATM) transports are available. The G_b interface assumes a Frame Relay transport. Other than these distinct Data Link–layer interfaces and some difference in QoS capabilities defined for GSM and UMTS, there is virtually no difference between the GPRS core for GSM and that for UMTS. The most notable functional allocation difference between GSM and UMTS is that the functions of header compression and ciphering of user data are in the core (and more precisely in the SGSN) for GPRS operating in G_b mode, and in the UTRAN for the I_u mode of operation. "G_b mode" is a way to identify the 3GPP PS core for GSM access only, and "I_u mode" is a way to identify the 3GPP PS core for UMTS access.

The GPRS Tunneling Protocol (GTP) is the protocol for the transfer of user-plane packets between the UMTS access and the UMTS core. GTP is also used over the G_n interfaces between SGSN and GGSN, and between SGSNs during handover. The G_n interface, unlike the I_u interface, uses GTP for both control and user planes (that is, both the GTP-C and GTP-U versions of GTP, specified in 3GPP TS 29.060 [181]). In the $I_{u\text{-}ps}$ interface, only GTP-U is used and the control is based on the RAN Application Part (RANAP) protocol.

In the course of 3GPP Release 7 development, an option to bypass the user plane of the SGSN under certain conditions (i.e., nonroaming user, no lawful interception activated, no CAMEL services enabled) has been introduced for the I_u mode of operation, thus enabling a direct tunnel between the access network and the GGSN. This feature seems to find its justification in higher data rates introduced in the UTRAN in 3GPP Rel-6 and Rel-7.

In the roaming case, the GGSN can be located in the home network or in the visited network, depending on the roaming agreements. This is different from the CS core, where the GMSC is always in the home network, regardless of the user location. When the GGSN is in the home network and the SGSN is in the visited network, the G_n interface connecting them becomes the G_p interface (which is only a difference in names, rather than a protocol-level difference).

Even though the capability to use a GGSN in the visited network is foreseen by the GPRS specifications, this option has not yet been used by operators as part of their roaming agreements. This is perhaps because of the practices linked to a legacy of operation of CS networks, but it is also because the operators want to have tight control of the user traffic at all times, so they wish to enforce policies or perform deep packet inspection in the home network when users are roaming.

The GGSN connects to external networks (also identified with the acronym PDN, or Packet Data Network) via the G_i interface, and the selection of an external network by the UE is done by submitting an access point name (APN) when the first Packet Data Protocol (PDP) context is created for a PDP session. A PDP context identifies a bearer used for a GPRS data session. For a single data session there can be one or multiple PDP contexts (up to 11 in UMTS), depending on the different levels of QoS used. Operator policies may determine the value of a default APN for the user when one is not specified at PDP context activation, and also the level of QoS allowed for a PDP context.

From 3GPP Release 7 onward, a framework to enforce QoS and charging policies on a per–IP flow basis has been introduced, so that the allowed QoS and the charging for each IP flow can be controlled based on the information derived from the applications associated with these flows. This framework is known as Policy Control and Charging (PCC), specified in 3GPP TS 23.203 [52]. As illustrated in Figure 4.5, PCC enables the

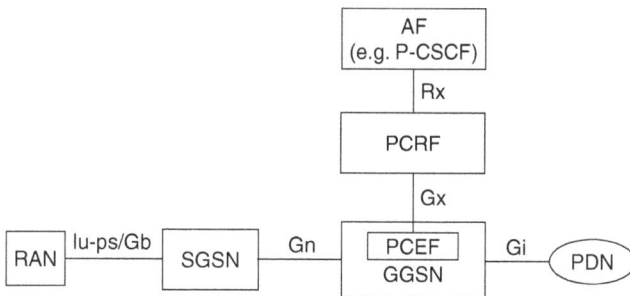

Figure 4.5 PCC framework

GGSN to act as the Policy and Charging Enforcement Function (PCEF) and interact with a Policy and Charging Rules Function (PCRF) via a DIAMETER-based G_x interface to enforce PCC decisions taken by the PCRF. The PCRF bases its decisions on the interactions with an Application Function (AF) via the DIAMETER-based R_x interface.

As VoIP becomes popular within wireline networks, it is likely that, given its acceptance, it will gain prominence in the cellular environment too. The IP Multimedia Subsystem has been defined to support VoIP service via the cellular PS domain in both 3GPP and 3GPP2. When VoIP is deployed in cellular, the PCC infrastructure is also introduced to support prioritization of voice flows (especially those related to emergency calls) and to provide fine-granularity QoS as well as gating of undesirable data that is not related to ongoing and accepted voice sessions. In the case of IMS, the Application Function is the Proxy CSCF (P-CSCF).

Radio Access Network and Terminal Aspects   Since aside from minor differences, GSM and UMTS share a common core, the substantial distinction between GSM and UMTS mobile systems is confined to the radio access network. Besides Physical-layer differences, where multiple access to the radio resources is based on TDMA for GSM and W-CDMA for UMTS (at least in its terrestrial aspects), the radio access networks for GSM and UMTS differ in both the respective network elements and their capabilities.

The GSM Radio Access Network   The GSM base station subsystem, depicted in Figure 4.6, includes a base station controller (BSC) connected via A-bis interfaces to a number of a base transceiver stations (BTSs). BTS supports radio frequency transmission to and reception from the mobile station (MS) via the U_m interface. The BSC, in addition to supporting inter-BTS mobility and encryption of voice communications, also performs transcoding between the voice-encoding format used over the radio interface and that used in the PSTN.

When the optional GPRS feature is added, the BSC supports the PCU (Packet Control Unit) used to add the packet data transmission capabilities without impact on the rest of the access network. When EDGE is deployed, theoretically raising the data rates to approximately 300 Kbps, then some modification to the BTS is needed.

Recently 3GPP has defined the I_u mode of operation for GSM, where the GSM BSS (depicted in Figure 4.6) is made compatible with the I_u interface used to access 3GPP core. However, since this would not change the overall GSM capabilities for the end user, which are limited by the BSS itself, and, especially, GSM terminal capabilities,

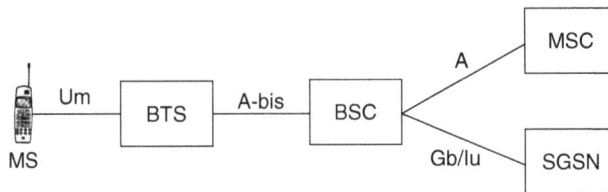

Figure 4.6   The GSM BSS (base station subsystem)

this mode of operation for the GSM BSS has not been implemented in real-life systems. Another recent set of enhancements is related to the potential need to support real-time services in GSM (to allow the reuse of the GSM BSS for VoIP applications), which also includes a packet handover capability (to allow shorter interruption time for real-time services when SGSN and BSS changes occur as the user moves).

GPRS Terminal Classes and Dual Transfer Mode (DTM)   As specified in 3GPP TS 23.060 [182], there are different classes of GPRS terminals, with the class indicating the mobile phone capabilities:

- **Class A**   These mobile phones can be attached to both GPRS and GSM services simultaneously.

- **Class B**   These mobile phones can be attached to both GPRS and GSM, using one of them at a time. Class B enables making or receiving a voice call, or sending/receiving an SMS during a GPRS connection. During voice calls or while sending SMS, GPRS services are suspended and then resumed automatically after the call or SMS session has ended.

- **Class C**   Mobile phones are attached to either a GPRS or GSM voice network; they cannot support two services simultaneously.

Since UMTS supports simultaneous PS and CS access, operators of GSM networks or mixed GSM/UMTS networks need to emulate the UMTS behavior in scenarios with patchy coverage (where data and voice sessions need to switch frequently between the two systems) and to provide subscribers with a consistent user experience whether they are camped on a 3G-capable network or not.

The definition of the GPRS class A mode of operation assumes a total independence between the CS and PS domains. This complicates the internal architecture of the terminal, so the optional capability known as Dual Transfer Mode (DTM) has been introduced, permitting the emulation of a class A terminal using DTM-capable terminals using *suspension of data transmission*. This is both a terminal and a network feature, so support on the terminal side alone is not sufficient for it to work.

The UMTS Terrestrial Radio Access Network (UTRAN)   The UMTS terrestrial radio access network (depicted in Figure 4.7) by design supports concurrent PS and CS access, and it has interfaces (named $I_{u\text{-}cs}$ and $I_{u\text{-}ps}$, respectively) to both the CS and PS core networks. These interfaces are supported by a radio network controller (RNC), which acts as a concentrator of traffic to and from many Node-Bs. (Node-B is the name of the network entity equivalent to a BTS in GSM.) The Node-Bs are connected to the RNC via an I_{ub} interface.

Since in CDMA technology a mobile terminal, or user equipment (UE) in UMTS terminology, can send to and receive from multiple base stations, the RNC combines and splits signals over the "active set" of base stations, in what is called the "Soft Handover" capability.

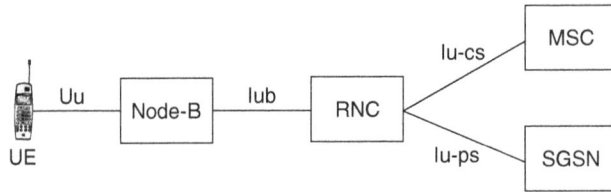

Figure 4.7 The UMTS terrestrial radio access network—UTRAN

3GPP2 Systems

In the family of 3GPP2 systems, we shall include the 2G and 3G CDMA systems, defined by ANSI/TIA and 3GPP2, respectively, despite the fact that the 2G system had been defined by TIA and not by 3GPP2 (like GSM was not defined by 3GPP, but by ETSI). The 3GPP2 system, also known as CDMA2000, is in fact an evolution of the cdmaOne system, known as the IS-95 family of CDMA technologies. These are defined in the TIA/EIA IS-95 specification, including the IS-95A [98] and IS-95B [99] revisions, which describe a complete mobile system. The world's first 3GPP2 3G commercial system was launched by SK Telecom in South Korea in October 2000 using CDMA2000 1X.

The IS-95A revision of IS-95 was published in 1995, and it was the one used for commercial 2G CDMA system deployments. It describes a wireless system based on 1.25 MHz CDMA channels, and it can provide voice services as well as circuit-switched data connections up to 14.4 Kbps. The IS-95B revision defines the support for *packet data services,* so it is sometimes categorized as a 2.5G system. IS-95B systems in fact offer 64 Kbps packet-switched data, in addition to voice services.

CDMA2000 technologies were defined by 3GPP2 and belong to a family of cellular wireless systems accepted as IMT-2000 technologies (i.e., 3G) by the ITU-T. This family includes

- CDMA2000 1x
- CDMA2000 1xEV-DO (Rev 0, Rev A, Rev B)
- Ultra Mobile Broadband, or UMB (previously known as CDMA2000 1xEV-DO Rev C), which represents an evolution toward 4G

CDMA2000 1x supports both circuit-switched voice communications and packet data speeds of up to 307 Kbps (as peak data rate). CDMA2000 1xEV-DO (Evolution-Data Optimized, or -Data Only) introduces a high-speed data broadband wireless network that can offer theoretical peak data rates beyond 2 Mbps in a mobile environment (in practice, the data rates consistently reach into 500 Kbps but rarely beyond that). CDMA2000 1xEV-DO is specified in IS-856, as the CDMA2000 High-Rate Packet Data Air Interface (also known as HRPD).

The 3GPP2 Circuit Core   The CDMA CS core network (used for IS-95 and CDMA 1x, as EV-DO is a PS-only system and relies on CDMA 1x to provide CS services) follows

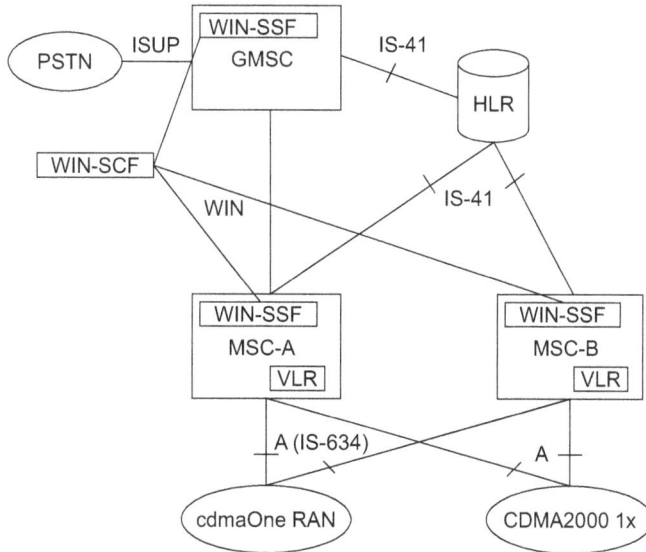

Figure 4.8 The CDMA CS core

similar principles to those of the GSM core, although there are protocol and main-stream implementation differences. The CDMA CS core in fact includes an MSC and a GMSC, as it is shown in Figure 4.8, and its interface to the RAN is also called the A interface; it follows the IS-634 [100] specifications.

Unlike in 3GPP systems, in CDMA, base station controller (BSC) and MSC are often implemented in a single physical node. In other cases, they communicate over a proprietary interface. The MAP protocol used in 3GPP systems to interface with the HLR is replaced by the TIA/EIA IS-41 [101]–specified protocol. Since this protocol needs to be used in roaming cases to access subscriber information for service authorization purposes, roaming between a CDMA network and a GSM/UMTS network is not possible without an interworking function between MAP and IS-41 (and of course the user needs a CDMA- and GSM/UMTS-capable phone). In addition, the flavor of the IN protocol (WIN) used in 3GPP2 is different from the CAMEL Application Part used in 3GPP.

The 3GPP2 Packet Core   Unlike the 3GPP and 3GPP2 circuit cores, which are essentially similar, the 3GPP2 packet core, shown in Figure 4.9, is quite different from the 3GPP core not only in its protocols, but also in its overall design. CDMA operators have chosen to adopt a packet core that is fully optimized for data, so they have used design principles more aligned with practices used by the Internet service providers. Therefore, they based their standardization efforts on reusing existing IETF protocols as they were, or promoted the enhancements of existing IETF protocols, rather than inventing their own.

Specifically, there was no need to define a special IP routing or tunneling setup (and its special and limited address ranges) for a walled roaming network like in GPRS,

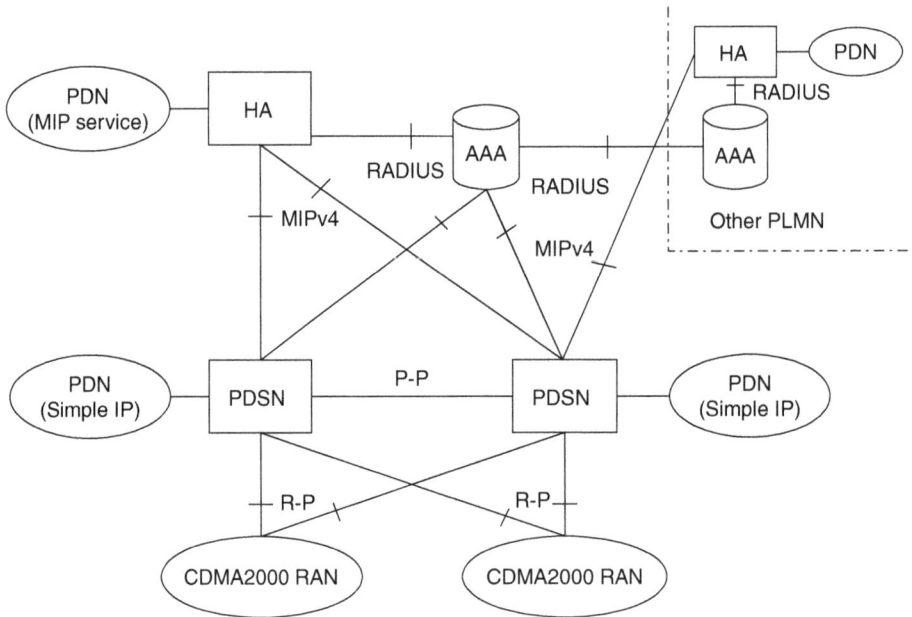

Figure 4.9 The CDMA2000 packet core

called GRX (GPRS Roaming Exchange), as roaming can work across the Internet using Internet Mobility augmented by the AAA support protocols that can withstand the security challenges encountered in the Internet.

Also, the HLR in CDMA networks is now used only for voice applications, as data subscribers' profiles are held in the AAA server, much like in all fixed IP networks. This approach is not aligned with the decision taken by 3GPP operators to keep storing subscribers' profiles for data in the HLR, or to evolve MAP to also support data applications. Similarly, there are no Intelligent Network–based services for the CDMA packet core, and all services and charging models are based on the use of AAA and non-IN-based frameworks.

The primary element in a CDMA packet core is the packet data serving node (PDSN). It is defined in the TSG-X X.S0011-001-D "cdma2000 Wireless IP Network Standard: Introduction" [108]. This node terminates PPP connections from CDMA terminals and performs user authentication for network access. It can also provide network access services and limited mobility within the scope of the BSCs directly attached to it via an R-P interface. The R-P interface is based on GRE encapsulation for the user plane and a protocol derived from Mobile IPv4 for the control plane. There is a homing relationship between PDSNs and CDMA2000 access network BSCs, so the PDSN's anchored mobility is limited geographically. The service provided to subscribers based on using a PDSN's anchored mobility only is called "Simple IP."

To achieve wider area mobility, that is, mobility between PDSNs, it is necessary for the terminal to support Mobile IP (MIP).[5] The network operator must deploy a MIP Home Agent (HA) and support a MIPv4 Foreign Agent (FA) for MIP4 or a MIPv6 access router for IPv6. Note that in real-life implementations the FA and access router functionality are typically supported in the PDSN platform. A MIP HA may be allocated to subscribers statically or dynamically. The latter is performed via AAA interaction during the Mobile IP registration phase. Simple IP and Mobile IP modes of operation are defined in X.S0011-002-D "cdma2000 Wireless IP Network Standard: Simple IP and Mobile IP Access Services" [109].

With the support of Mobile IP and a relatively technology-neutral approach to data services support, CDMA networks could be poised to be transformed into converged multiaccess technology networks in a more straightforward way than their GSM/UMTS counterparts.

Radio Access Network and Terminal Aspects   Let's start the discussion of the CDMA systems RAN with the analysis of the data rates and services capabilities supported by various flavors of CDMA 3G systems.

CDMA2000 1xEV-DO Rev 0 supports bidirectional peak data rates of up to 153 Kbps and an average of 60–100 Kbps in commercial networks in a 1.25 MHz channel. Release 1 can deliver peak data rates of up to 307 Kbps. CDMA2000 1x handsets are backward compatible with cdmaOne systems, so dual-mode capability is not needed for a commercial terminal to use both 2G and 3G, unlike in 3GPP technologies.

CDMA2000 1xEV-DO Rev 0 offers peak data rates of up to 2.4 Mbps in the "forward link" (also known as downlink in 3GPP) and 153 Kbps in the reverse link (or uplink in 3GPP parlance), in a single 1.25 MHz FDD carrier. As usual, this is to be checked with reality, and in most cases in commercial networks, CDMA2000 1xEV-DO Rev 0 may deliver average (or sustained) throughput in the range of 300–700 Kbps in the forward link and 70–90 Kbps in the reverse link.[6]

To cope with initially spotty coverage of CDMA2000 1xEV-DO, CDMA2000 1xEV-DO devices include a CDMA2000 1x modem to be compatible with CDMA2000 1x and cdmaOne systems (it should not be forgotten that a CDMA 1x device is also compatible with cdmaOne).

CDMA2000 Evolution   CDMA2000 1xEV-DO Revision A (also known as DOrA) is an evolution of CDMA2000 1xEV-DO Rev 0 providing higher peak data rates, offering more symmetric performance between forward and reverse links, and, most important, supporting QoS levels compatible with delay-sensitive multimedia applications, including VoIP and streaming video. Rev A supports 3.1 Mbps in the forward link and 1.8 Mbps in the reverse link in a 1.25 MHz FDD carrier.

---

[5] We invite those willing to probe further on Mobile IP details to check out a fundamental book on the subject by James Solomon, *Mobile IP, The Internet Unplugged* [165].

[6] Source: CDMA Development Group (CDG).

The first commercial deployments of DOrA systems took place in the course of the year 2006, at the same time as High-Speed Downlink Packet Access (HSDPA). Note that HSDPA enhanced with improved uplink performance, known as HSUPA, was not yet deployed at the time. In commercial networks, Rev A average (or sustained) throughput is in the range of 450–800 Kbps in the forward link and 300–400 Kbps in the reverse link[7] with typical latency as low as 50 ms. Multicast capabilities have also been added to DOrA by means of OFDM technology to enable multicast content delivery.

The CDMA2000 1xEV-DO Rev A capability to support VoIP makes it an ideal technology for converged networks where voice and other multimedia services such as video telephony and push-to-talk are supported over a single all-IP infrastructure. Improved reverse link speeds allow users to send and receive large amounts of data in ways previously possible only in fixed networks or over Wi-Fi.

DOrA networks support existing CDMA EV-DO Rev 0 applications and devices. This backward compatibility preserves an operator's network investments. Multimode devices, however, are needed to access CDMA 1x and cdmaOne. CDMA2000 1xEV-DO Revision B [110], defined in 3GPP2 C.S0024-B, evolves Rev A by aggregating multiple DOrA 1.25 MHz channels to provide higher performance for multimedia services, bidirectional data transmissions, and VoIP-based services.

Rev B Peak data rates are proportional to the number of carriers aggregated (Rev B is a "multicarrier" version of Rev A). So, in principle, if a 20 MHz spectrum bandwidth is available, fifteen 1.25 MHz Rev A channels can be combined, and peak data rates of 46.5 Mbps in the forward link and 27 Mbps in the reverse link could be achieved. By using a 64-QAM (Quadrature Amplitude Modulation) modulation scheme, forward link peak data rates of up to 4.9 Mbps per 1.25 MHz carrier are achievable, which means fifteen carriers can deliver up to 73.5 Mbps in the forward link direction. It should be noted that in these multicarrier setups, the 1.25 MHz carriers do not have to be adjacent to one another, giving operators significant flexibility in their deployments. Device compatibility considerations are similar to those illustrated for Rev A.

UMB   UMB is yet another step on the way of CDMA evolution. UMB specs are being completed by 3GPP2 at the time of this writing. UMB promises to deliver peak data rates of up to 280 Mbps on the forward link and 68 Mbps on the reverse link; a very low latency below 20 ms; and scalable, noncontiguous, and dynamic channel (bandwidth) allocations of 1.25 MHz, 5 MHz, 10 MHz, and 20 MHz.

The CDMA radio access network architecture differs depending on whether a plain IS-95 and CDMA2000 1x access (see Figure 4.10) or a CDMA2000 1x EV-DO access is used. It should be remembered from the discussion so far that the only CDMA access networks supporting CS service are CDMA2000 1x and IS-95, and EV-DO is an enhancement that leads to data-only operation and to the possibility to support real-time communications over a PS domain (in Rev A).

---

[7] Source: CDMA Development Group.

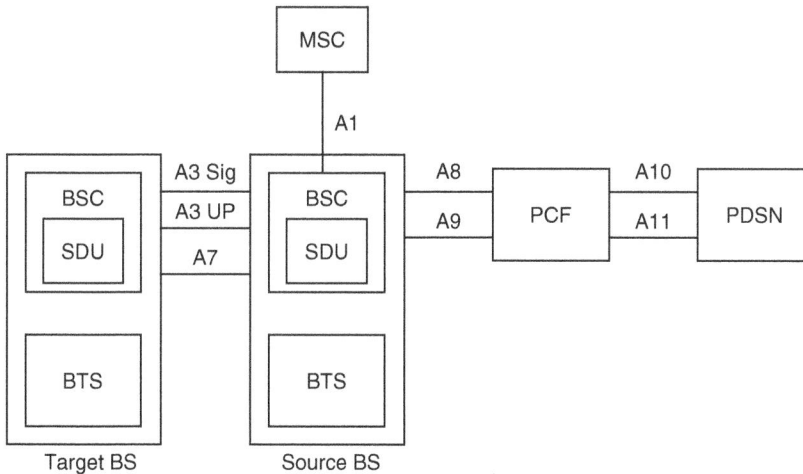

Figure 4.10 The CDMA access network

CDMA RAN   The CDMA RAN for CS services only is made of a base station, in most commercial implementations comprising a base station controller (BSC) and a base transceiver station (BTS). The standards do not define an open interface between the BTS and the BSC, although they may be based on different physical nodes. The BSC supports functions such as the SDU (selection distribution unit) necessary to support CDMA soft handover (requiring the BSC to perform frame selection and distribution to an active set of base stations).

To support packet data in CDMA, the RAN is augmented with a packet control function (PCF). The PCF is normally implemented as a part of the same physical platform as the BSC. The session management and mobility management functions are normally located in the core network, but the option to place these functions in the PCF has also been introduced in the course of the HRPD specification. In the latter case, the interworking with an MSC supporting this function in the CDMA 1x network requires an IWS (interworking solution) module in the PCF. This option, depicted in Figure 4.11, is specified in the 3GPP2 A.S0009-A "Interoperability Specification (IOS) for High-Rate Packet Data (HRPD) Radio Access Network Interfaces with Session Control in the Packet Control Function" [95].

The RAN architecture also evolves to support the concept of IP-based RAN and more efficient distribution of functions between the BSC and the BS. One example is the placement of the scheduling function in HRPD. Up to CDMA 1x, the CDMA RAN is based on a BSC connecting to a base station using some form of TDM transport (T1s). The introduction of scheduling in the base station makes it possible for IP-based RAN to be deployed for CDMA 1x EV-DO networks, and thus benefit from lower backhaul costs and improved distribution of network elements in the network. This also reduces OPEX for operators, as they can reduce the number of BSC hosting locations.

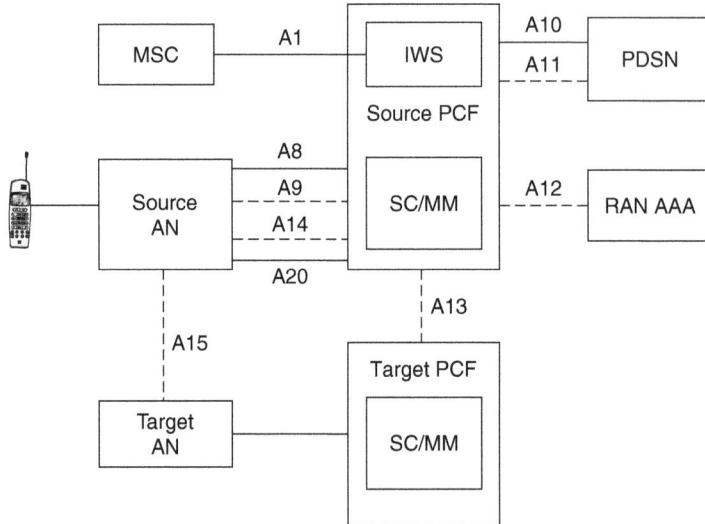

Figure 4.11   CDMA access network for packet data with Session Control in the packet control function

The Need for VCC   Devices supporting 1x and Rev A cannot use both systems at the same time. When VoIP is rolled out in areas of Rev A coverage, the deployment of Rev A is unlikely to be ubiquitous, so when a voice call is started in an area of the RAN where Rev A is supported, it may be the case that due to varying radio conditions or limited Rev A coverage, the call will need to be "handed down" to 1x circuit-based voice and vice versa.

This requirement has lead to the definition of a Voice Call Continuity specification for the handover of voice calls between HRPD and CDMA 1x (that is, packet- and circuit-based systems), based on the assumption that the two technologies cannot be accessed at the same time. The system requirements for the support of this mechanism are defined in 3GPP2 S.R0108-0 "HRPD-cdma2000 1x Interoperability for Voice and Data System Requirements" [112], and the specification of the actual handover procedures is in 3GPP2 C.S0075-0, "Interworking Specification for cdma2000 1x and High-Rate Packet Data Systems" [183].

A Look into the Future

As 3G cellular systems have been defined, deployed, and enhanced, the 3GPP and 3GPP2 communities are looking ahead to shape the next generation of mobile systems. In doing so, they are targeting convergence of the standards by evolving 3GPP and 3GPP2 systems in the same direction. In large part, this direction is driven by some prominent 3GPP2 operators interested in getting access to the larger embedded base of 3GPP subscribers and realizing better economies of scale.

The evolution of the 3GPP system is being investigated as part of the System Architecture Evolution (SAE) work for the architectural aspects, and in the Long Term

Evolution (LTE) project for the radio access network aspects. The resulting system will be a PS-only system (no circuit services will be supported) delivering data rates in the order of 100 Mbps in the downlink and 50 Mbps in the uplink. The new radio transmission technology, known commonly as LTE after the project name or E-UTRA, is supported by the evolved UTRAN (E-UTRAN).

The 3GPP Evolved Packet System   As defined in the standards, the E-UTRAN will be substantially simplified and will consist of a single element called *Evolved NodeB (E-NodeB)*. The E-NodeB element, roughly speaking, groups all the functions once supported by the UMTS Node B and RNC. Like the UMTS RNC, the E-NodeB supports the encryption and header compression functions for the user plane. The interface between the E-NodeB and the core network, equivalent to the $I_{u\text{-}ps}$ interface in UMTS, is based on the GTP protocol for the user plane and the *S1* application protocol for the control part.

The 3GPP Evolved Packet Core network (EPC) is made by a control-plane entity called the *mobility management entity (MME),* whose functions include paging and mobility management. As the MME is responsible for the signaling interface with the UE, it also supports the relevant security functions, including authentication and signaling traffic encryption. The user plane is handled by two entities: the serving gateway (SGW) and the PDN gateway (PGW). The overall architecture, documented in 3G TS 23.401 [120] and 3G TS 23.402 [121], is depicted at a high level in Figure 4.12 (the references can provide more detail).

The SGW concentrates the S1 interface user plane to E-NodeBs, buffers data in the downlink, and triggers paging when the UE is in idle mode. The SGW may also anchor mobility toward legacy UMTS and GSM SGSNs. The SGW interacts with the MME via an open interface. There is a single SGW per UE at any given time, and there is

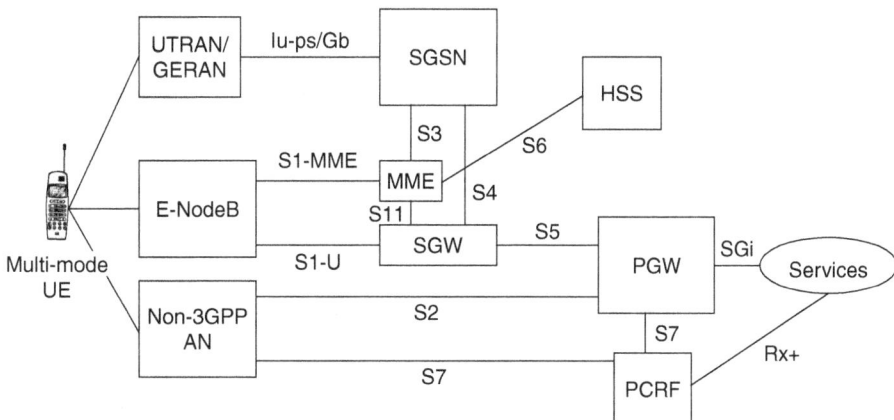

Figure 4.12 High-level SAE architecture

no geographical homing relationship between an E-NodeB and an SGW, improving reliability by avoiding a single point of failure in the system.

The PGW provides access to packet data networks. Potentially a UE may access multiple PDNs concurrently via one or multiple PGWs through the SGW. The PGW also supports a Mobile IP HA functionality to provide for network-based (Proxy-MIP [117]) or client-based (Mobile IP [118]) IP mobility support with non-3GPP access networks over the S2 interface.

Support of mobility across a variety of access technologies, documented in 3G TS 23.402 [121], is one of the most important features of the evolved system, and it is also among the most controversial ones. In fact, the deployment of IETF-based IP mobility in the 3GPP system to support non-3GPP access may lead to the eventual introduction of MIP-based roaming interfaces instead of or in addition to the GTP-based mobility support and roaming interfaces that have been so far typical for 3GPP systems. Changing roaming interfaces will of course imply operations changes as a consequence, a fact that, for obvious reasons, is not being accepted well by many operators, so there has been some friction on this topic.

The 3GPP2 System Evolution   As far as 3GPP2 is concerned, its system evolution has identified a UMB radio interface (Ultra Mobile Broadband), and there is a great deal of cooperation with 3GPP in order to share the core network architecture and to define interworking between LTE and CDMA systems, thus placing 3GPP2 into an evolutionary path toward the same system architecture (and potentially also the same radio interface) as 3GPP.

WiMAX: A Migration Away from Traditional Cellular Systems

Unlike the majority of the cellular systems reviewed in previous sections, WiMAX is a mobile system designed to deliver only packet data services. WiMAX is based on the IEEE 802.16 [114] radio interface, which can operate in both licensed and unlicensed spectrum bands. As with cellular systems, it is possible to use WiMAX terminals equipped with RUID, USIM, or SIM cards. However, in contrast to cellular, which requires a purchase of a smart card or a telephone as part of the subscription, the WiMAX model also allows a casual setup of network access, similar to today's Wi-Fi hotspots. Average bit rates advertised of WiMAX systems are up to 70 Mbps, but of course, this figure needs to be regarded with some caution, as in practical deployments these may be significantly lower (in the order of 10 Mbps average data rate).

WiMAX was not born as a mobile system per se; it was thought of as more of a wireless alternative to cable and DSL for fixed broadband applications and the evolution of Wi-Fi. But when a wide area wireless system is available, the temptation to also make it usable for mobile applications is irresistible. For this very reason, the IEEE was prompted to define various flavors of IEEE 802.16.

The original IEEE 802.16 (also known as IEEE 802.16a) specifies a radio interface operating in the 10 to 66 GHz range. The IEEE 802.16d evolution of the standard added the operation in the 2 to 11 GHz bands. Its subsequent update, IEEE 802.16e, defines a radio interface suitable for mobile applications.

Let us also clarify the distinction between different modes of WiMAX operation. The terms "fixed WiMAX," "mobile WiMAX," "802.16d," and "802.16e" are frequently used. This is what they mean:

- **802.16d**   This standard is more precisely identified as IEEE 802.16-2004 [115]. However, in the industry this is mostly known as 802.16d.

- **Fixed WiMAX**   WiMAX systems use IEEE 802.16d as radio interface.

- **802.16e**   Technically speaking, what is commonly known as 802.16e is an amendment to IEEE 802.16-2004. Its most accurate name would be IEEE 802.16e-2005 [116]. This standard supports full mobility, unlike 802.16d.

- **Mobile WiMAX**   WiMAX systems use IEEE 802.16e as a radio interface; hence these systems enable mobility (although a mobile WiMAX system can also be used for Fixed Wireless Broadband Access applications).

Standardization and Deployment

The WiMAX Forum is the standards forum that has specified the WiMAX system. Release 1 enables, among other things, the support of interworking with 3GPP and 3GPP2 networks, at the level of providing access to the Internet based on sharing credentials (loosely coupled interworking), but no real handover capability. This and other advanced capabilities are, however, in the scope of the work of the WiMAX Forum.

One of the biggest challenges for WiMAX is the ability to be deployed in a regulated spectrum in a way that does not create a global lack of device-network interoperability (or better, compatibility) for roamers. In fact, regulators seem to give permission to use WiMAX in quite different bands across the globe. For instance, in the U.S., companies like Sprint, Nextel, and Clearwire own the spectrum around 2.5 GHz. In other regions deployments are foreseen in the 2.3/2.5 GHz (the dominant frequency expected in Asia), 3.5 GHz, or 5 GHz band. In the European Union it has also been proposed to allocate some spectrum (yet undefined at the time of writing) for noncellular wireless communications technologies such as WiMAX. Clearly, this situation will force device manufacturers to produce terminals compatible with a large number of operating frequencies, potentially even more than current quad-band GSM handsets support. This may drive complexity and cost in terminals, which is undesirable, as this tends to impact mass market acceptance.

Like cellular systems, the WiMAX system (depicted in Figure 4.13) defines, broadly speaking, an Access part also known as Access Service Network (ASN) and a Core part referred to as Connectivity Service Network (CSN), although some of the classic cellular core functions, such as accounting and user access control, are supported in the WiMAX Access part.

The WiMAX Connectivity Service Network   The Connectivity Service Network depicted in Figure 4.14 serves as an interconnection point between the ASN and services networks (e.g., application service providers, the Internet, the operator's own services, etc.). Also, the CSN makes it possible for a subscriber to move between ASNs and

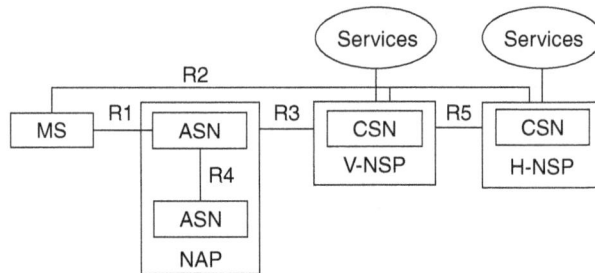

Figure 4.13 The WiMAX high-level architecture

to roam, by providing the necessary anchor point for mobility and the AAA mechanisms necessary to support roaming.

A CSN may in fact act as a AAA proxy toward the home network CSN, with the ASN playing the role of the AAA client, as shown in Figure 4.14. The CSN includes the MIP Home Agent terminating Layer-3 mobility support tunnels, whether they are based on PMIP [117] or client MIP defined in [118] and [119]. The AAA server in the CSN is used to determine whether a user has access rights to the wireless service, and it also may act as a AAA proxy toward the home network or toward customer/services networks connected to the CSN.

The AAA server stores WiMAX subscription profiles (including static QoS profiles) and performs accounting data collection. When dynamic QoS policies are supported, the CSN may include a policy decision point (e.g., a PCRF), but this is not defined in the WiMAX Release 1.0. The enforcement of QoS policies may take place in the ASN.

The visited network is connected to the home network via the R5 interface, and the visited ASN connects to the visited CSN via the R3 interface. It should be noted that the HA may be located in the V-CSN for local breakout of traffic or in the H-CSN for the home-routed

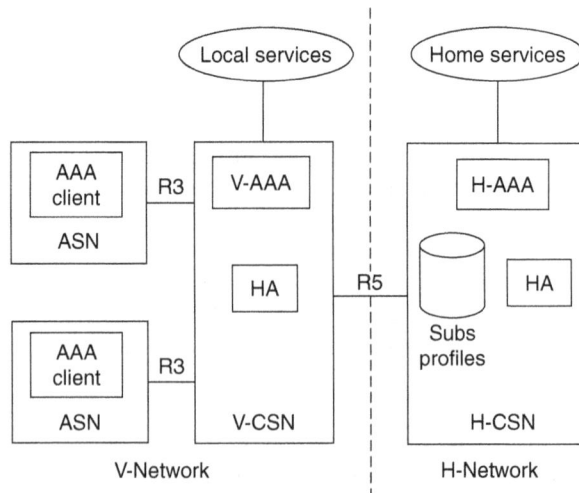

Figure 4.14 The CSN in the WiMAX mobile system

traffic roaming mode. The CSN, in its role of interfacing with the services network, also performs IP address allocation functions and network access authentication.

The WiMAX Access Service Network   The WiMAX Access Service Network is composed of two network elements: a base station (BS) and an ASN gateway (also abbreviated as ASN GW henceforth), as shown in Figure 4.15. The functions in the BS and the ASN differ depending on which of the three WiMAX profiles a vendor complies with. In fact, the WiMAX Forum has defined the ASN as a set of functions that can be grouped differently depending on the profile being used. The compliance to a profile ensures that there is a known way to interoperate with nodes that are built according to it.

The ASN gateway may act as the anchor for inter-BS mobility, and support control-plane functions to coordinate with other entities in the ASN itself, the MS, and the CSN. The MS is associated with a single ASN GW; but the BS to which MS is attached can use multiple ASN GWs (for redundancy, to avoid single points of failure, and for load-sharing reasons). Figure 4.15 shows a typical ASN architecture. Different ASNs or different ASN GWs may interact via the R4 interface. The R3 interface connects the ASN GW to one or more CSNs.

An ASN GW may also include a decision point for Dynamic QoS policies and an enforcement point. This decision point may be separate from the enforcement point on the R7 open interface. The BS connects to one or more ASN GWs via the R6 interface; the R8 interface between BSs exists for handover purposes.

WiMAX Profiles   The WiMAX standard framework includes three different usage and deployment profiles for a WiMAX system: A, B, and C. These profiles are characterized by different allocations of functions to BS and ASN GW.

The main characteristics of profile A are these:

- Handover control is in the ASN GW.

- Radio Resources Management (RRM) is in the ASN GW, so that it is a centralized function.

- Intra-ASN mobility is based on open interfaces (the open interfaces R6 and R4 are both used).

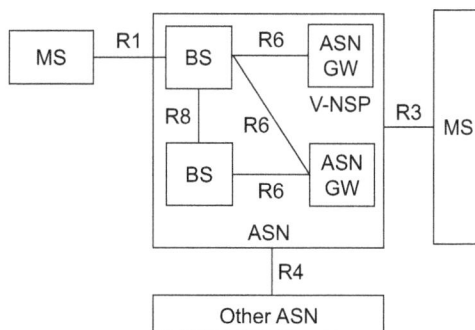

Figure 4.15   The ASN in the WiMAX mobile system

For profile B, *Intra-ASN* mobility is based on proprietary mechanisms, but *Inter-ASN* mobility uses the R4 interfaces. No allocation of functions to ASN GW or BS is specified, due to proprietary nature of the ASN.

For profile C, these are the essential features:

- Handover control is in the base station.

- RRC (Radio Resource Control) is in the BS (more specifically, the RRM in the BS). An "RRC Relay" is in the ASN GW, to relay the RRM messages sent from BS to BS via R6.

- Intra-ASN mobility is based on open interfaces (the open interfaces R6 and R4 are both used).

Table 4.1 summarizes the main characteristics of each profile.

QoS   From a QoS perspective, the Release 1.0 specification of the WiMAX system defines the following potential capabilities:

- Preprovisioned service flow creation, modification, and deletion
- Initial service flow creation, modification, and deletion

Table 4.1   WiMAX Profiles and Their Function Allocations to ASN GW and BS

Profile	ASN GW Functions	BS Functions
Profile A	Authenticator Key Distributor Data Path function (user-plane handling) HO function Context server/client MIP Foreign Agent (MIPv4) MIP Access Router (MIPv6) Paging controller RRM Service flow authorization (QoS policy decisions)	Auth Relay Key Receiver Data Path function (user-plane handling) HO function Context server/client Paging agent Radio Resource Agent Service flow management (QoS policy enforcement)
Profile B	Not specified	Not specified
Profile C	Authenticator Key Distributor Data Path function (user-plane handling) HO function Context server/client MIP Foreign Agent (MIPv4) MIP Access Router (MIPv6) Paging controller RR relay Service flow authorization (QoS policy decisions)	Auth Relay Key Receiver Data Path function (user-plane handling) HO function Context server/client Paging agent Radio Resource Agent Radio Resource Control Service flow management (QoS policy enforcement)

- QoS policy provisioning between AAA and SFA
- Service flow ID management

The scope of Release 1.0 is limited to preprovisioned service flows and to IEEE 802.16 radio aspects only of QoS, with no definition of end-to-end behavior. The characteristics of the QoS service over the WiMAX radio are the following:

- Connection-oriented service
- Five QoS services at the air interface, namely:
  - UGS (Unsolicited Grant Service)—for Real Time Constant bit rate service
  - RT-VR (Real Time—Variable Rate)
  - ERT-VR (Extended Real Time—Variable Rate)
  - NRT-VR (Non-Real Time—Variable Rate)
  - BE (Best Effort)
- Provisioned QoS parameters for each subscriber based on a subscription profile
- Policy-based admission of service flow requests

A subscription profile for a user is defined as a number of service flows (which are quintuples {Source IP Address, Destination IP Address, Source Port, Destination Port, Protocol}), each of which is associated with the QoS parameters. This information is provisioned in a policy server (e.g., in an AAA subsystem). With the static service model, a WiMAX terminal is not allowed to change the parameters of provisioned service flows or create new service flows (which instead are both possible within the scope of the dynamic service model). The dynamic service flow creation is triggered by the terminal or the applications accessible via the WiMAX system.

Cellular Friend or Foe: Voice over Wi-Fi

So here we have it. Cellular is synonymous with wireless access and mobile voice communications. Indeed, up until recently that was the only technology supporting truly mobile communications. (Mobile WiMAX technology provides standard support for voice communications; however, at the time of this book's writing, no broad-scale WiMAX voice commercial deployments have taken place.)

Nowadays, however, new technologies such as Wi-Fi, originally designed to be mere extensions of wired Ethernet to provide wireless data network access, are being quickly adopted for mobile voice. With the proliferations of public hotspots and the rise of convergence technologies such as FMC, Voice over Wi-Fi is quickly maturing into a credible complement, if not an alternative, to traditional cellular.

In this section we provide a brief overview of this novel approach to the support of wireless voice service, and analyze the impact of Voice over Wi-Fi (VoWi-Fi) on both the end user and the service provider.

Technology Fundamentals

Wireless local area networking (WLANs) or Wi-Fi[8] (for Wireless Fidelity) has truly taken the world by storm in the recent years. This phenomenon coincided (albeit independently) with the rise of Voice over IP as one of the primary alternatives to traditional fixed PSTN voice communications. So it did not take long for the users and the industry to realize that the two are even more useful when combined into a service offering potentially capable of ending consumer dependency on both traditional PSTN and cellular telephony.

While in theory combining VoIP with Wi-Fi to make it mobile looked like a natural extension of fixed VoIP solutions, early real-world implementations proved to be difficult. This happened mainly because wireless networks based on IEEE 802.11 [126] (a group of standards underlying Wi-Fi) were originally prone to interference, did not support voice prioritization and guaranteed QoS, were not optimized to carry real-time traffic, and so forth (as seen in Figure 4.16). So, to address these deficiencies, the industry had to turn to short-term proprietary implementations while at the same time mounting the concerted standardization efforts needed to make Wi-Fi more voice friendly.

Wi-Fi Standardization   Wi-Fi is currently in widespread use all around the world as both replacement and transport media complementary to wired LANs. Indeed, 802.11-based WLANs are essentially a wireless extension of the wired Ethernet LANs standardized

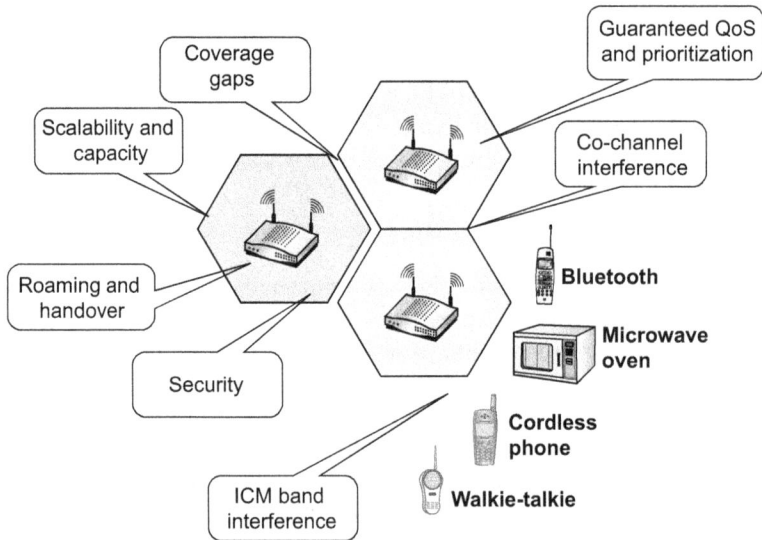

Figure 4.16 VoWi-Fi challenges

---

[8] We are going to use both terms interchangeably throughout the text for the purposes of discussion.

in IEEE Standard 802.3 [127]. Wi-Fi is another example of wireless technology originally designed to only carry data (similarly to WiMAX). Unlike WiMAX, however, WLANs were not intended to be deployed in wide areas but rather to serve the needs of local, not highly, mobile users. Therefore, original Wi-Fi specifications did not include roaming and handover support.

IEEE adapted standard 802.11[9]—which provided the foundation for Wireless LAN technology—in 1997. The standard was consequently revised in 1999 and continues to be enhanced with amendments and supplements by the same standards body. Following the introduction of 802.11, IEEE later coined a more popular term: WLAN. Independently of IEEE, the Wi-Fi Alliance, a trade organization specifying and testing commercial equipment for interoperability and compliance with IEEE standards, came up with the term Wi-Fi, which soon became a favorite with popular media (presumably for its "catchiness").

The latest version of the standard was approved by IEEE in March 2007 under the name IEEE 802.11-2007, which now combines eight previous amendments: a, b, d, e, g, h, i, and j.

Wi-Fi Components   Unlike cellular systems, Wi-Fi technology and networks[10] are not very complex, and their components are fairly simple and inexpensive, which was one of the decisive factors in the world-wide proliferation of Wi-Fi just a few short years after its introduction. A Wi-Fi network requires a client device with a wireless *network interface card (NIC)* and an *access point (AP)* terminating a radio link to multiple clients and connecting the wireless LAN with the wired infrastructure.

While a variety of NICs continue to be available in all shapes and sizes from laptop PC cards to flash and USB dungles, the majority of commercial implementations nowadays incorporate them into mobile devices themselves in the form of a chipset combined with an RF subsystem with a dedicated antenna.

The APs are following a similar trend. While dedicated access points are still widely available, one can more often find them supporting additional functionality such as switching or routing and sold as multifunction devices or even bundled with NICs and other equipment such as desktop PCs and laptops. Strictly speaking, two distinct types of APs are available today: regular APs, also called "fat," or intelligent and simplified ones called "thin." The thin APs are most often deployed in combination with aggregating Wi-Fi switches also called WLAN controllers in large-scale enterprises, or in metro or campus installations.

Further, when used for VoIP traffic, Wi-Fi customer-premises equipment typically must perform even more duties. It must support routing and Wi-Fi networking and also provide support for both VoIP and PSTN telephony via terminal adapter capability for customer legacy equipment. Therefore it is not uncommon to find all three functions embedded into one device. Such an integrated device might be designed to support

---

[9] After the number of engineers assigned to design it.

[10] Note the important distinction.

two or more service set identifiers (SSIDs), one for VoIP traffic and one for regular data communications. The purpose of this approach is to provide service separation and enable QoS and prioritization solutions as described in the following sections. Figure 4.17 lists the examples of commercial AP implementations and their properties.

While we are at it, let's define SSID. It is a text string of up to 32 characters identifying a common access point domain in a Wi-Fi network. All clients intending to communicate with a particular AP must be programmed with its unique SSID. The original thinking behind the SSID was to use it as an additional security measure. This approach, however, does not provide real security, since even if the SSID is not broadcast by the AP, it can be easily obtained, e.g., by snooping using any of the widely available hardware or software protocol analyzers.

Radio Interface   WLAN ranges specified in IEEE standards can reach as far as 300 meters outdoors under perfect conditions with no obstacles or interference present. With the use of amplification devices such as directional antenna arrays, the signals can reach up to 1 kilometer, often even in the presence of obstacles. The distance and obstacles, however, significantly affect the signal strength, which in most cases decreases exponentially with the distance. Once a client device has connected to an 802.11 access point, it makes the data rate determination based upon the available signal strength and sometimes other parameters such as QoS profiles and power-saving policies.

The spectrum used by Wi-Fi falls into the unlicensed Industrial, Scientific, and Medical (ISM) band category (2.4 to 2.5 and 5 GHz in the U.S. and Europe), and therefore does not require federal licenses in most countries to be used by individuals or offered commercially by service providers. Many other devices such as cordless phones, microwave ovens, Bluetooth devices, and headsets are permitted to emit radiation in this spectrum, which creates the potential for interference (see the accompanying sidebar).

Each Wi-Fi AP is assigned to a channel. That channel consists of frequencies in the 2.4 GHz, 2.5 GHz, or 5 GHz range of the radio spectrum, depending on the specific flavor of the 802.11 standard being used. For example, in the U.S. there are 11 different

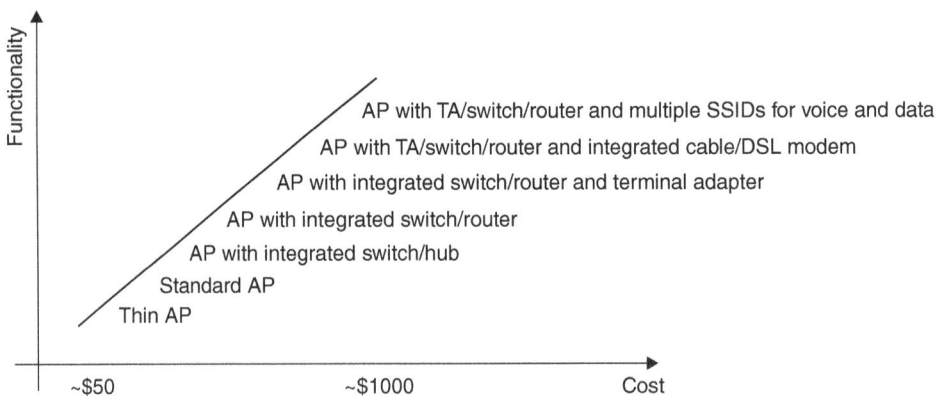

Figure 4.17 Wi-Fi AP types

but overlapping[11] channels available in 802.11b/g wireless networks (see the letter nomenclature description in the next section). In other regions of the world such as Europe and Japan, the standard supports 13 and 14 channels, respectively. Similar to its wired predecessor, Ethernet, the Wi-Fi channel access is based on the Carrier Sense Multiple Access–Collision *Avoidance* (CSMA-CA)[12] technique.

Interference

The 2.4 GHz ISM band reserved for free public use spans the frequency range 2400–2483.5 MHz. The devices operating in this band include two types: *unlicensed* such as microwave ovens, Bluetooth devices, and certain cordless phones, and *licensed* such as amateur radio and RF remote controls. Note that according to regulation (such as those by the FCC, the regulatory body in the U.S.), unlicensed devices are not allowed to interfere with the licensed ones.

A number of cordless telephones operate in the 2.4 GHz ISM bands using a randomly determined set of frequencies, or "frequency hopping." Potentially they can interfere with Wi-Fi traffic, but this does not happen often. For one thing, the user would have to be in an active data session and engaged in a phone conversation over a cordless set at the same time. Using VoWi-Fi phones simultaneously with analog PSTN ones can potentially exacerbate the problem, but the chance that a household or business would keep two types of phones, PSTN and Wi-Fi cordless, is slim.

Bluetooth devices are another potential source of WLAN interference. However, the transmission power of the Bluetooth devices is an order of magnitude lower than that of the cordless phones, so their potential for interference only arises in close proximity to WLAN devices (at ranges of 1–3 meters). Also, the Bluetooth specification starting from version 1.2 specifies the adaptive frequency hopping (AFH) method to avoid interference. The designers of dual-mode Wi-Fi/cellular phones, which typically support Bluetooth, and other Wi-Fi devices likely to experience interference have also created a number of proprietary solutions providing effective workarounds.

Finally, home microwave ovens radiate a narrowband signal between 2450 and 2460 MHz, creating interference potential. Most home microwave ovens, however, are well shielded, and experiments have shown that they would only interfere with an 802.11 network if the AP were within a few feet of one of the 802.11 endpoints.

---

[11] Channels 1, 6, and 11 were designated as nonoverlapping channels with additional spectrum provided for better separation.

[12] This protocol has its roots in the other standard created at the University of Hawaii in 1970 called ALOHA-NET, which was one of the most important milestones in data networking as we know it today. Note that the protocol underlying Ethernet is really called CSMA-CD for collision detection. The schema had been modified for the wireless environment to provide collision avoidance.

The networks defined by 802.11 standards originally relied on two types of radio frequency modulation techniques: frequency-hopping spread spectrum (FHSS) and direct-sequence spread spectrum (DSSS). While one of the 802.11 implementations called 802.11a was based on FHSS, most of the commercial solutions that followed (starting with 802.11b) used DSSS due to its higher tolerance to interference and potential to support higher bit rates.

Architecture   The 802.11 standard defines two modes of operation for WLANs:

- *Ad hoc* mode, also known as independent basic service set (IBSS), where the clients communicate with each other
- *Infrastructure* mode, where an AP provides client access to a network

A WLAN in ad hoc mode essentially functions as a peer-to-peer network that does not require servers. It allows two or more clients with NICs to communicate directly with each other. An example of an ad hoc use case might be a group of visitors at a meeting with their laptops connected to each other, creating a mesh separate from a corporate network of their host.

The infrastructure mode is a more widespread architecture. Its topology includes one or more clients connected through APs to the IP core infrastructure. The architecture of Wi-Fi networks operating in an infrastructure mode consisting of one AP is referred to in standards as a basic service set (BSS). Multiple BSSs compose an extended service set, or ESS.

From the high-level view, functionality of APs is similar to that of a cellular network base station. In fact, multiple APs forming an ESS in enterprise, campus, or metro mesh setups are treated similarly to cells in a typical distributed wireless cellular system, as Figure 4.18 illustrates. One of the important distinctions between cellular *systems* and Wi-Fi *networks* is that the original 802.11a, b, and g standards did not support IP-layer mobility or the ability to maintain an active data session while changing access points and subnets[13] and many other functions necessary to provide wide area service.

802.11 Variations   The initial success of Wi-Fi quickly resulted in considerable standards activity, which produced both numerous variations of the original 802.11 standard specifying different bands or data rates and extensions dealing with functionality omitted from the original specifications, such as QoS, inter-AP handoff, security, and others. What follows is a quick list of some of these standards and extensions to the original standards identified by lowercase letters following the 802.11 identification in alphabetical order. As mentioned in the earlier section "Wi-Fi Standardization," extensions a through j have recently been merged under the 802.11-2007 version of the original 802.11 standard.

---

[13] This capability is specified by the 802.11r extension discussed later in the chapter.

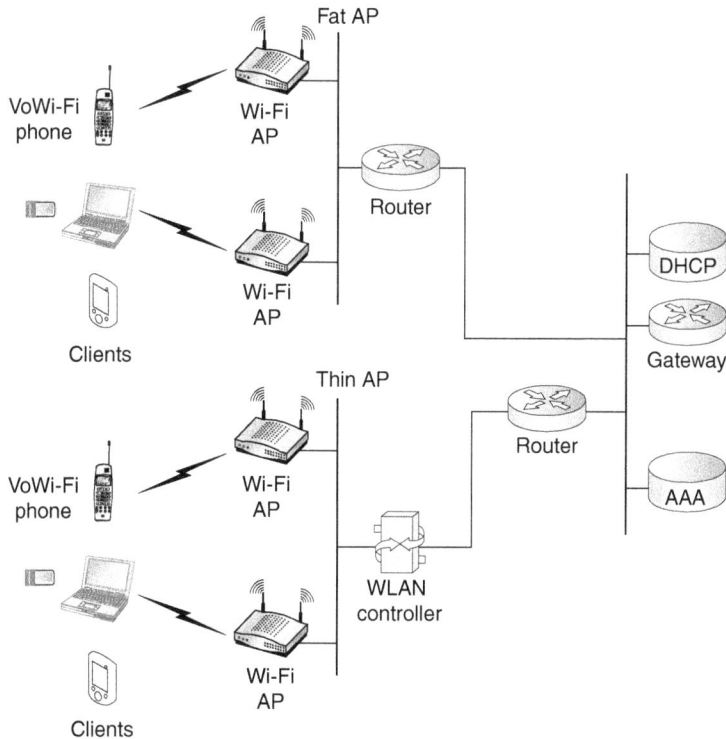

Figure 4.18 WLAN ESS example

802.11a   The 802.11a [131] standard, along with 802.11b, was the first amendment to the original 802.11 specification. This standard defines operation in the 5 GHz band with 300 MHz of bandwidth. The theoretical maximum bit rate of IEEE 802.11a is 54 Mbps. The 802.11a devices operate in the 5 GHz frequency band using OFDM technology. The reason for selecting the 5 GHz band was the belief that the 2.4 spectrum may become congested.

By design, 802.11a supports 24 nonoverlapping channels (though only 8 can be used at any given time), which is a significant improvement over the b and g standards described next. The more channels are offered, the more options are available for the user, making it easier to avoid interference. The trade-off for the increased bandwidth with 802.11a and the move up the frequency ladder is a range typically limited to about 50 meters, which is roughly half of that provided by 802.11b. The limited range of 802.11a was partially responsible for its relatively cold reception by the enterprise market it was originally targeting.

802.11b   The 802.11b [132] standard was the second of the two initial amendments to the original 802.11 specification. The 802.11b standard specifies data rates of 1, 2, 5.5, and 11 Mbps in the 2.4 GHz spectrum and is based on DSSS. This standard is one of the most widely adopted in today's commercial products both in residential and enterprise environments.

802.11d   The 802.11d [138] standard was developed by IEEE to extend the original 802.11 specification to countries where the original specification is not applicable or allowed due to unique spectrum allocations and other regulatory reasons or, in the standard words, "additional regulatory domains."

802.11e   The 802.11e [139] standard specifies a set of QoS enhancements of the original 802.11 standards prompted mostly by low applicability of the standard Wi-Fi for VoIP and other types of real-time traffic. The 802.11e standard supports differentiation for various types of data on the network such as voice, video, and other multimedia and more delay-tolerant types of communications.

On the footsteps of 802.11e, the Wi-Fi alliance introduced the Wi-Fi Multimedia (WMM) certification used to ensure interoperability of 802.11e-compliant equipment. WMM defines four traffic categories (or access categories in WMM terminology): voice, video, best effort, and background.

802.11g   The 802.11g [133] standard defines a higher bit rate alternative to 802.11b. The higher bit rates are achieved by transmission on the same 2.4 GHz radio frequency band used by 802.11b. Like 802.11a, 802.11g is also based on the OFDM technology and supports a maximum theoretical bit rate of 54 Mbps. 802.11g is backward compatible with 802.11b.

For backward compatibility with 802.11b devices, the standard specifies support of complementary code keying (CCK) modulation based on the RC4 cryptographic algorithm and providing both access authentication and encryption of traffic. The standards allow mixed b and g systems; however, g speeds will degrade significantly as b clients are associated with combined b/g access points.

802.11h   The use of the 5 GHz frequency band in Europe caused some problems with regulatory requirements and interference with other services. To overcome these problems, an amended version of IEEE 802.11a, IEEE Standard 802.11h [134], was developed. The 802.11h standard also introduced some advanced techniques such as Transmit Power Control (TPC), potentialy reducing transition power by 50 percent, and Dynamic Frequency Selection (DFS), providing automatic channel hopping to avoid interference or overload. In Norway and a number of other European countries, the 802.11a/h systems can only be used indoors due to local government regulations.

802.11i   Originally wireless LAN security was based on SSID and—fast becoming legacy—the Wired Equivalent Privacy (WEP) security scheme. WEP security is weak and vulnerable to eavesdropping because of its basic keying scheme and poor vector initialization. The IEEE 802.11i [136] security specification has addressed WEP shortcomings by introducing a new security framework superseding the original WEP and enabling robust authentication, encryption, and key rotation.

The Wi-Fi Alliance commercialized selected portions of the 802.11i standard as WPA, which stands for Wi-Fi Protected Access. The 802.11i standard is also referred to as WPA2 by the Wi-Fi Alliance. Instead of the weak RC4 cipher, the 802.11i standard relies on the stronger Advanced Encryption Standard (AES) cipher.

802.11n   In response to the demand for even greater throughput, IEEE has began working on the 802.11n standard, specifying data rates of up to 248 MBps, and backward compatible with both 802.11b and 802.11g. 802.11n defined the elaborate MIMO-based antenna schema supporting separate antenna arrays for sending and receiving signals (similar to that specified by 802.16 standards) to significantly improve data rates through *spatial multiplexing* and ranges through *spatial diversity*. The 802.11n standard supports both 5 GHz and 2.4 GHz frequencies. Work on the 802.11n definition is still ongoing at the time of writing, with the current approval target set for the end of 2008.

802.11r   The 802.11r standard—which is currently under development by IEEE—also known as *fast BSS transitions* or *fast roaming,* is an 802.11 extension specifying fast data handover between access points. The standard is especially applicable in enterprise or metro environments where there is a need to preserve the data session continuity while moving between multiple APs. It is expected that VoIP will eventually become one of the standard's main applications; however, for this to happen, IEEE must address real-time traffic delay concerns.

The WLAN supporting the 802.11r standard will in many respects behave similarly to today's cellular network (at least in terms of data handover). Interestingly, 802.11r is mostly focused on the handover setup and not the actual handover handling, which is currently left for vendor interpretation.

Addressing VoWi-Fi Challenges   VoWi-Fi deployment presents a number of unique challenges to all players in its ecosystem. Some of the challenges (as shown earlier in Figure 4.16) include:

- Support for QoS and prioritization of voice in the environment originally created to carry data traffic

- Support for strong security mechanisms (especially in enterprise applications)

- Support for fast roaming and seamless in-call handover in wide area and metro deployments

- Slow introduction and limited selection of mass-market VoWi-Fi-capable mobile devices

- High VoWi-Fi device power consumption

- Low access point capacity and limited coverage

New 802.11 standards and extensions as well as numerous proprietary approaches are being successfully applied to address these challenges, enabling VoWi-Fi solutions, originally relegated to vertical markets, to go mainstream. The examples include residential and small office systems complementing broadband VoIP services and replacing cordless telephony, enterprise solutions extending VoIP PBXs, and FMC solutions where VoWi-Fi is seamlessly combined with cellular—the focus of this book.

Next we explore how these challenges are being addressed.

QoS   In general, running Voice over IP, especially in the enterprise and hotspot "WAN" environment (in other words in places where there are multiple APs and multiple streams of data), presents many challenges for a Wi-Fi network. The most critical among these is achieving and uniformly maintaining acceptable audio quality, for example, by minimizing network delay in a mixed voice and data environment.

Wired or wireless, 802.11 networks were originally not designed for real-time streaming media or support for guaranteed packet delivery rates. Therefore, congestion on the wireless network or drop in throughput, without traffic differentiation, can quickly make any kind of voice transmission unintelligible or at least severely degrade the user experience. When voice service is provided over the IP protocol running over a wireless interface, prioritization and quality of service become hard requirements, mainly because of the variable and finite throughput of the air link, which is further affected by interference, constantly changing medium characteristics, and varying distance to the AP.

Conducting real-time communication, such as voice or multimedia streaming, dependably is especially difficult in *public* Wi-Fi networks using ordinary "best effort" operation. Excessive latency in such environments can be caused by either one-way packet delay or extensive buffering needed to address jitter (variance of the packet arrival time). Latency exceeding 200–300 milliseconds (depending on the codec, device, and specific type of voice communication used) during a two-way conversation starts negatively affecting the user experience and quickly degrades the conversation. The original Wi-Fi standards were designed to operate using statistical multiplexing of user traffic contending for access to the air interface based on a "best effort" approach. That means that the quality of a particular type of service cannot be guaranteed, especially when the channel utilization is increasing, which in turn causes an incremental rise in packet collision and retransmission rates.

Guaranteed QoS[14] mechanisms combined with prioritization schemas were introduced to Wi-Fi in the 802.11e standard, primarily addressing latency, jitter, and error rate.

The 802.11e standard provides four QoS traffic categories:

- Voice
- Video
- Best-effort
- Background

---

[14] QoS is based on the idea that transmission rates, error rates, and other characteristics can be measured, controlled, and to some extent, guaranteed in advance.

802.11e relies on the Hybrid Coordination Function (HCF). HCF is a single-channel access protocol enhancing the original 802.11 Distributed Coordination Function (DCF) used by CSMA-CA for medium allocation and channel access coordination. HCF identifies and prioritizes different types of traffic by introducing a concept of Traffic Classes (TC) and applies the Controlled Channel Access (CCA) protocol to CBR traffic to ensure a constant bit rate above the minimum quality threshold set for a particular traffic type.

802.11e HCF introduces two main options for controlling channel access:[15]

- Enhanced Distributed Channel Access (EDCA)
- Hybrid-Coordinated Controlled Channel Access (HCCA)

EDCA   The Enhanced Distributed Channel Access schema prioritizes traffic classes (higher-priority traffic sent first) by assigning each traffic class a Transmit Opportunity tag (TXOP), which identifies the period of time during which the client is allowed to transmit an unlimited amount of data.

HCCA   The HCF-Coordinated Controlled Channel Access (HCCA) schema is used when a more precise QoS definition (than that provided by EDCA) is needed. HCCA also allows for the *reservation* of TXOPs with the AP and defines traffic streams (TSs) in addition to traffic classes. HCCA introduces a concept of Controlled Access Phase (CAP)—a period initiated by an AP for contentionless communication with a mobile station.

Along with ensuring consistent quality of service, the HCCA also provides a robust call admission control (CAC) mechanism to make sure the allocated channel bandwidth is sufficient to carry a given number of simultaneous voice conversations. CAC enables the APs to calculate the available bandwidth, make handover decisions (if other APs in the range have unused capacity), and throttle the traffic accordingly.

QoS Summary   Both the HCCA and EDCA options have their pros and cons. The HCCA mechanism is more appropriate for setting up fine-grained QoS policies—allowing for precise definition of acceptable latency and jitter limitations—and therefore may be better suited for commercial VoIP and multimedia services. However, HCCA is more complex to implement than EDCA and to date has not seen much support in commercial Wi-Fi products, which almost uniformly support EDCA.

Needless to say, in addition to the approaches defined in 802.11e, all the common IP-layer QoS standards such as MPLS, DiffServ,[16] and RSVP (described in Chapter 2) can also be used in conjunction with techniques offered by 802.11e to improve overall QoS of VoWi-Fi system traffic.

---

[15] Interested readers are welcome to explore further in the original IEEE standard specification and the book by Frank Ohrtman, *Voice over 802.11* (Artech, 2004).

[16] DiffServ is an IP-layer QoS framework defined by IETF that takes the IP type of service (TOS) field, renames it in the Differentiated Services Code Point (DSCP) field, and uses it to carry information about IP packet service requirements.

Security   The security support is as important for voice as for data traffic to ensure conversation privacy, enforce authentication and authorization, and preserve overall network integrity. However, the demands of strong security on a wireless network, which for example may inadvertently increase latency or jitter, can have a direct impact on voice quality. Security support becomes even more difficult in a Wi-Fi environment supporting fast roaming, as the properly secured network must still be able to unobtrusively support mobility (both roaming and handover).

All 802.11 specifications, such as a, b, and g, support a simple WEP security mechanism. By design, WEP may support a 40-bit key and a 104-bit key combined with a 24-bit initialization vector, resulting in 64- and 128-bit encryption, respectively. WEP is based on RC4[17] cryptography and provides both access authentication and encryption of traffic. WEP, however, allows for transmission of clear text (original data) and cipher text (encrypted data) in the open during session establishment between client and AP. This makes it easy to extract a shared secret key from the traffic between the AP and the client for anyone with a sufficiently sophisticated protocol analyzer.

Further, WEP is usually implemented with manual distribution of static keys shared between all clients associated with an AP. This deficiency in particular makes it clear why WEP-based security is not suitable for enterprise installations with multiple clients because of the need to change keys on all associated clients if one of them is lost.

The 802.11i standard, referred to as WPA2 by Wi-Fi Alliance, was developed to address most if not all of the original WEP shortcomings. The main security improvements introduced by 802.11i include the support for 802.1x/EAP (Extensible Authentication Protocol, providing robust authentication and authorization), along with advanced encryption algorithms such as the Temporal Key Integrity Protocol (TKIP) and the Advanced Encryption Standard–Counter Mode/CBC-MAC Protocol (AES-CCMP), designed to improve confidentially protection. 802.11i also provides mechanisms designed specifically to aid fast roaming, such as preauthentication.

The 802.11i standard proposes two authentication methods:

- *Personal,* which is a basic option based on preshared keys intended for residential use

- *Enterprise,* which is a more advanced option developed for enterprises and, potentially, certain metro deployments

While the 802.11i personal option essentially enhances WEP by eliminating key transmission and requiring instead distribution of master shared secret keys to users through other means, the "Enterprise" option provides a completely new security approach based on centralized authentication and dynamic key distribution.

---

[17] Named incidentally for Route Coloniale 4, a road in Vietnam and a famous battle of the first Vietnam war, which took place in 1950.

The 802.11i Enterprise option is based on 802.1x and EAP IEEE standards, which provide RADIUS-based mutual client/server authentication, support centralized policy definition, and offer dynamic distribution of encryption keys. The session time-out in the Enterprise option triggers reauthentication and new encryption key generation, which further enhance privacy protection.

Authentication   The EAP/802.1x model is based on three main objects, depicted in Figure 4.19, involved in the authentication process:

- An *Authentication server* such as RADIUS or DIAMETER
- A *Supplicant,* a client supported by a device that needs to be authenticated
- An *Authenticator,* an element acting as a proxy of an authentication server, typically an AP

In this model the specific authentication method is defined by EAP between the Supplicant and the Authentication server. To gain access to a network, the Supplicant (client) must go through a two-step process of initial association with the server followed by mutual authentication between the client and the server. After the authentication is performed by the Authentication server, the Authenticator (AP) grants or disallows client network access based on the authentication process results. If the authentication is successful, the client and server originate the same encryption key, which is never transmitted in the open. When the encryption key is generated, the Authentication server distributes a session key to the Authenticator, which it uses to encrypt the broadcast key used to encrypt the session.

The solutions implementing EAP typically support login credentials based on *two-factor* authentication, requiring the user to supply an ID and a password for initial association.

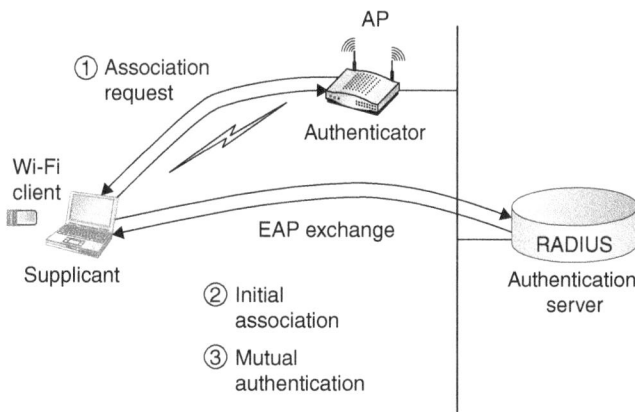

Figure 4.19   802.11i authentication model

Since EAP leaves many aspects of actual protocol execution up to the vendor, the industry came up with several commercial implementations called EAP types. Popular examples of EAP types include:

- EAP–Transport Layer Security (EAP-TLS), defined in IETF RFC 2716 [140]
- EAP–Subscriber Identity Module (EAP-SIM), described in IETF RFC 4186 [84] and used in 3GPP-defined solutions relying on SIM cards
- EAP-LEAP (for Lightweight EAP), a proprietary mechanism developed by Cisco Systems supporting dynamic WEP keys and mutual authentication between client and Authentication server
- Protected EAP (PEAP), a proprietary mechanism developed by a consortium of vendors, which provides client authentication but does not support encryption

EAP-TLS enhances the original EAP model with the support for digital certificates. When the client attempts association with the server, the Authentication server first sends the client a certificate for validation. Following a successful validation, the client sends its own certificate back to the server, which is validated in the same way. Upon successful mutual authentication the *EAP-Success* message is sent to the client, the Authenticator (AP) is notified, and the normal EAP procedures resulting in generation of dynamic WEP key take place.

The EAP-SIM option, based on the challenge-response mechanism, was developed in part with converged Wi-Fi/3GPP cellular solutions in mind. Unlike EAP-TLS, EAP-SIM requires the mobile device to support a SIM card à la GSM and also requires the Authentication server to be connected with the GSM core (specifically AuC and MSC), because this EAP implementation uses the SIM card and International Mobile Subscriber Identity (IMSI—a number uniquely associated with a particular SIM) to supply authentication credentials to identify the user. The AuC function in GSM allows the MSC to authenticate SIMs upon initial connection to the network. The AuC also generates an encryption key to protect session privacy and a random number called a triplet (a 64-bit random number, a 32-bit response [SRES], and a 64-bit K_c key), which is used as a shared secret between the SIM and the AuC.

The authentication process starts with the Authentication server requesting GSM triplets from the AuC and returning the random numbers with a checksum (derived using SRES and K_c components of the triplet) to the client with SIM as a challenge. Using the random number provided by the server, the client calculates the checksum using the device's SIM card to generate its own SRES and K_c numbers. The resulting checksum is then compared to the one received from the server; if they match, the authentication is successful and the client initiates a challenge to the server using the same procedure. Upon successful mutual authentication, the EAP-Success message is sent to the Authenticator and a sufficiently robust key is generated.

Despite its complexity and reliance on the access to AuC in the GSM core infrastructure, EAP-SIM is particularly well suited for use in dual-mode FMC handsets. On the other hand, it is not extendable to CDMA FMC solutions or any other systems with non-SIM-based handsets.

Confidentiality Protection   Along with addressing the problem of weak authentication, one of the goals of 802.11i was to improve the original WEP encryption. This is achieved via the support of two encryption options: the Temporal Key Integrity Protocol (TKIP) enhancing the WEP RC4 cipher, and the Advanced Encryption Standard (AES) combined with the Counter Mode with Cipher Block Chaining Message Authentication Code Protocol (CCMP), providing more-advanced encryption and ciphering.

The TKIP enhancements include per-packet keying (PPK), a message integrity check (MIC), and extension of the original WEP initialization vector from 24 bits to 48 bits. The TKIP PPK mechanism supports generation of different unicast per-packet keys to allow multiple initialization vectors to use different keys (as opposed to WEP). The PPK is also used to provide the broadcast key rotation to protect broadcast and multicast WLAN traffic against the same threats.

Mobility   To be considered for carrier-scale commercial and enterprise-wide deployments, VoWi-Fi solutions must support mobility to perform on a par with cellular systems. The mobility support requirements include:

- **Macromobility**   Roaming and active call handover between Wi-Fi and other wireless access technologies such as cellular
- **Micromobility**   Roaming and active call handover between Wi-Fi APs within the same subnet, in different subnets of the same network, or in different Wi-Fi networks

Clearly, both types of mobility must be supported in the environments with both multiple APs, where the VoWi-Fi mobile station is expected to change its physical location and roam between them, and multiple access networks such as Wi-Fi, WiMAX, cellular, and others. Examples of such environments include wide areas with mixed networks and VoWi-Fi-only locales such as corporate offices, college campuses, and metro installations, potentially capable of covering whole cities or conceivably even regions. VoWi-Fi micromobility, however, is less relevant in residential or SoHo installations, where VoWi-Fi technology is used for a cordless telephony or in combination with cellular for residential FMC services.

The majority of today's general-purpose access points are not equipped to support inter-AP roaming and handoffs. Wi-Fi technology was originally designed with the data users in mind and did not account for typical telephony use cases (walk or drive and talk vs. walk or drive and type messages or browse the Web).

To support micromobility, the mobile station must preserve its IP address to avoid reattachment to the network, and it must always perform fast reauthentication to protect overall network security and integrity. These requirements present additional challenges to handover timing, which must not exceed approximately 250 milliseconds (a threshold also set for cellular systems) to avoid choppy voice transmission negatively affecting the user experience.

In the section that follows we are going to analyze the technology allowing the support of IP-layer micromobility, while deferring discussion of macromobility to Chapter 5.

Inter-AP Handover   The standardization of Wi-Fi roaming is still ongoing in IEEE "task group r," with the 802.11r standard expected to be ratified in 2008. The 802.11r specification defines fast BSS (basic service set) transitions for seamless handover for the a, b, g, and upcoming n standards. The issues currently under consideration by the working group include both VoIP and data mobility, minimization of handover delay, and compatibility with 802.11n and 802.11i (robust authentication). The 802.11r standard is based on the make-before-break scheme, in which the security association, QoS profile, and other connection properties are established with the neighboring *target* "to" AP before leaving the *serving* "from" AP.

With the 802.11r standard still in the works, in recent years the industry has come up with multiple proprietary roaming solutions designed to satisfy the requirement of the systems with multiple APs and provide a user experience on a par with that of cellular system users. The majority of such architectures are based on two types of access points:

- A thin AP/WLAN switching network architecture with the control consolidated in the core infrastructure components

- An intelligent AP network architecture in which most of control is concentrated in "smart" or "fat" APs themselves

Inter-AP roaming must be supported so that the handovers can take place on the OSI Data Link layer when the client roams between APs within the same subnet or on the Network (IP) layer using Mobile IP when the client roams between APs in different subnets or even networks (see Figure 4.20).

Typically, inter-AP handoff is a multistep process beginning when the mobile station monitors the RF link quality of the serving AP and evaluates it against other APs within range. The RF link condition deteriorating below a certain threshold will trigger handover based on a conditional algorithm. Along with RF monitoring, the mobile station also periodically scans the network for foreign APs to determine potential availability for roaming.

The process of handover itself starts with reauthenticating with the target AP. This procedure depends on a particular security protocol in use, and if not implemented properly, may have a significant impact on overall handover latency. After the mobile station is reauthenticated and granted permission to connect to a new AP, the transfer of QoS context occurs and the mobile device is associated with a new AP. In the process of handover involving multiple subnets or networks, the IP address of a mobile station must be kept constant, which can be accomplished utilizing protocols like Mobile IP.

Dealing with Handover Latency   Keeping handover latency sufficiently low to make it imperceptible to the user is a significant technical challenge with the majority of the secure Wi-Fi authentication methods. Two of the standardized approaches to speeding up the authentication process defined in the 802.11i standard are based on the reuse of the previous authentication results for generating new encryption keys or modifying the results of the earlier authentication to derive the new keys to avoid full reauthentication.

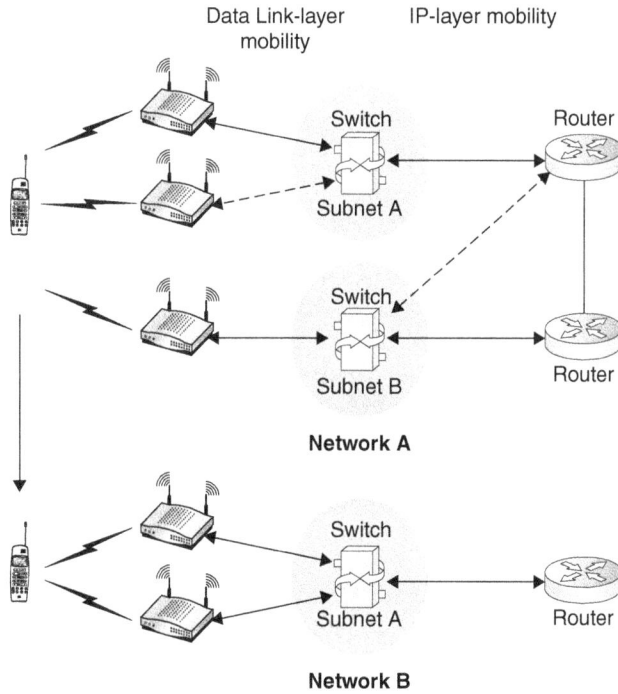

Figure 4.20   Wi-Fi micromobility

The first method is called *preauthentication*. This method requires the serving AP to be connected to potential target APs over the wired network. With such a connection in place, a target AP can "preauthenticate" the client prior to the actual move, so during the handover only minimal 802.1x/EAP authentication procedures will be required.

The second method is called pair-wise master key (PMK) caching. When a client associates with a particular AP supporting PMK using one of the EAP procedures, it caches the EAP encryption key. If the mobile station roams away from the AP and then "comes back," it will be able to reuse the cached key to avoid full reauthentication. The PMK flavor called opportunistic PMK, developed for use in an 802.11i enterprise security framework, is applied in thin AP architectures, allowing caching of an encryption key in a switch or AP controller.

Along with reducing reauthentication times, other proprietary implementations and IEEE draft standards are focusing on reducing the time needed for neighboring AP scans. One popular scheme enables a broadcast of the list of neighboring APs in the beacon. This reduces the scan time for the clients and saves battery life. Another scheme proposes deploying APs in preset channel numbers only, thereby limiting the scan channel list for the clients.

Handover Triggers   In Wi-Fi networking, the client decision to initiate handover is usually triggered by reaching certain threshold conditions. These are examples of such conditions:

- The radio link quality falls below an acceptable threshold.
- The maximum retry count is exceeded.
- The data rate falls below an acceptable threshold.
- The serving AP becomes overloaded.

When one of these triggers fires, the mobile station initiates the scan of available 802.11 channels. On each discovered channel, the client station sends a probe and waits for probe responses or beacons from APs on that channel. The probe responses and beacons received from APs are discarded unless they have matching SSID and encryption settings. Thus qualified, remaining APs are then prioritized based on a set of predetermined parameters such as availability, throughput, or proximity. After the target AP is determined, the handover procedure is initiated.

Reducing Client Power Consumption   The original Wi-Fi specification was written for data-centric mobile stations such as laptops with a relatively long battery life. When Wi-Fi is used for voice communication, the mobile devices are usually handheld with a form factor similar to that of a cordless phone, or even a cellular phone in the case of dual-mode FMC solutions. Smaller devices mean smaller batteries and consequently degraded active and standby time. In fact, early technology trials yielded Wi-Fi standby times roughly half those of cellular, a fact that was pretty much guaranteed to disappoint users.

To address these shortcomings along with proprietary approaches, two power-saving solutions were introduced in the 802.11e standard:

- Unsolicited Automatic Power Save Delivery (U-APSD)
- Scheduled Automatic Power Save Delivery (S-APSD)

In U-APSD, the client sends a "standby ready" notification to the AP before going on standby, which causes the AP to start backing up the data destined for the client and switch to "beacon" mode. The AP in beacon mode sends periodic beacons to the dormant client indicating if there is any data available. The client in a dormant state would periodically "wake up" only to read the beacons. Depending on the beacon type, the client will either go back to a dormant state or switch to an active state.

The S-APSD scheme provides a similar power-saving mechanism for clients operating in the HCCA mode. This scheme was designed specifically for real-time traffic such as VoIP. It goes a step farther than U-APSD by allowing the clients to go on standby for preprovisioned periods, without having to listen for the beacons.

Access Point Capacity   The individual AP capability to support multiple simultaneous VoIP streams is an important measure of overall VoWi-Fi system capacity, which is especially significant in enterprise and hotspot environments. Hotspots installed in public spaces like libraries, cafés, and hotel lobbies are expected to support significant numbers of simultaneous conversations. In enterprises, the same requirements are applicable to APs serving factory floors, conference rooms, reception areas, and the like.

The capability of 802.11b systems to support VoIP was first analyzed back in 2004 in a paper by Hole and Tobagi [175]. The paper provided theoretical calculations of the maximum possible number of VoIP sessions that could be supported by a single AP. Figure 4.21 shows the dependence of the number of sessions on data rate under three different sets of conditions:

- First (leftmost in each set) is a theoretical analysis assuming perfect RF conditions and no delay.

- Second (middle of each set) is a theoretical analysis assuming perfect RF conditions with a one-way Wi-Fi link delay of 10 ms or less.

- Third (rightmost in each set) is a theoretical analysis assuming the bit error rate (BER) of the radio channel to be 1.0×10^{-4} with a delay of less than 10 ms.

Not surprisingly, the number of supported sessions was shown to increase with the data rate. Supporting paper observations, later practical experience showed that, for example, a single 802.11b AP can support between 8 and 12 simultaneous VoIP sessions in an environment without interference. This number, however, can quickly drop to 2–3 with the introduction of interference or increasing distance between the AP and clients. The 802.11a or g solutions can increase this rate to 15–20 (albeit, at the expense of coverage radius). It was also observed that the AP throughput itself may not be as much of a limiting factor as latency, prioritization, and jitter.

Regulatory: E911, CALEA, and Other Mandates   A commercial VoWi-Fi service must comply with the emergency calling framework whenever possible, behaving on a par with today's cellular and fixed VoIP systems. Currently, emergency calls in a cellular network, governed by E911 in the U.S., E112 in the EU, and E110/119 in Japan, are routed

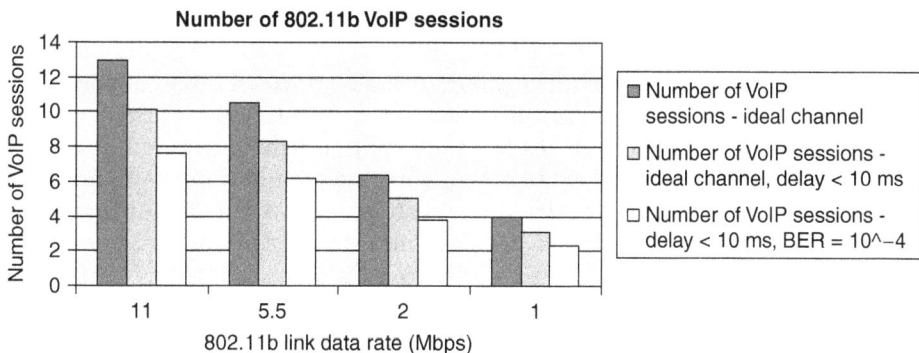

Figure 4.21   Access point capacity (source: Hole and Tobagy)

to the appropriate Public Safety Access Point (PSAP) using the routing digits received from the location services platform.

In a VoIP and hence in a commercial VoWi-Fi solutions, in most countries, emergency calling support is required by the government to deliver the caller's location or address of the subscriber to the PSAP along with the callback number. This can be accomplished by appropriately routing VoIP calls from media gateways in service provider networks or establishing IP connections directly to appropriately equipped PSAPs.

For example, in the U.S. the National Emergency Number Association (NENA) is currently evaluating a number of methods for bringing VoIP 911 calling into E911 systems that provide Selective Routing and Automatic Location Information (ALI). NENA is likely to recommend that Voice over IP providers use newly available interfaces in the E911 systems for customers that have fixed locations and are using telephone numbers from their local area code.

It must be noted that standard PSTN emergency calling approaches are often not applicable in commercial VoIP and, hence, VoWi-Fi implementations. Since VoIP service is access independent, the subscribers may change their address or even use several temporary points of attachment to the network during the course of the contract. The VoWi-Fi service users that can travel between hotspots within the same city or country are even harder to track; albeit during admission to Wi-Fi service, the identity and location of the Wi-Fi hotspot serving the user can be easily passed on to the PSAP via the Authentication infrastructure.

Convergence of Mobile Systems

At the end of this chapter we are going to take a quick look at convergence of mobile systems, which should provide a good segue to the discussion of convergence of fixed and mobile systems.

One of the most important aspects of the next-generation wireless communication systems is the capability to support simultaneous access to CS and PS services. This allows for more flexibility in circuit voice and packet data services creation (parallelism in access to service), and also to the transition from one technology, supporting circuit voice, to another, supporting packet-based transmission, using the methods like VCC (first mentioned in the section "3GPP Systems" of this chapter and described in detail in Chapter 5) defined in 3GPP Release 7. In addition, Mobile IP can support the *make before break* handover on the IP layer when it is possible to access both the source technology and the target technology simultaneously, when both support PS services.

Also, the ability to concurrently support access to CS and PS services via different access technologies minimizes the interruption time during an active voice call even when the network is not cooperating with the terminal in the handover process. This property is particularly desirable in systems with a significant installed base of both CS and PS technologies (such as Wi-Fi and circuit cellular systems), which need to be converged while minimizing the impact on the infrastructure. It is also possible to use DTM to support simultaneous access to GSM CS services and GPRS services, or to use UMTS to provide simultaneous access to its PS domain and its CS domain.

Table 4.2   Most Likely Terminal Operation (SR = single radio, DR = dual radio)

	GSM	UMTS	LTE	WiFi	WiMAX	CDMA
GSM	N/A	SR	SR	DR	DR	SR[1]
UMTS		N/A	SR	DR	DR	SR[2]
LTE			N/A	DR	SR	SR
Wi-Fi				N/A	SR	DR
WiMAX					N/A	DR
CDMA						N/A

[1] Applicable to provide GSM or CDMA customers with a terminal capable of global roaming support.

[2] Applicable to provide UMTS or CDMA customers with a terminal capable of global roaming support, but also in countries like Korea, where there are operators with both these access networks.

Often, to minimize the cost of multimode terminals, they are built to support only one cellular technology in an active transmission state at a time (that is, other supported technologies cannot be active at the same time—in the approach known as a "single radio terminal" (SR)—when one is in an active transmission state). In this case, the support of seamless mobility or change of domain requires cooperation by the network in preparing the handover. While it is always possible to implement "dual-radio" (DR) terminals, it is expected that single radio will be prevalent in many commercial implementations. Table 4.2 presents what the authors consider the most likely scenarios for dual-mode terminals. The case of multimode (e.g., tri-mode UMTS, E-UTRAN, Wi-Fi) terminals could be inferred by a logical combination of the relevant dual-mode options.

It turns out that in most cases the single-radio strategy applies when multiple cellular access technologies are used, while Wi-Fi and WiMAX matched with a cellular access would normally require a dual-radio operation (for instance to allow a user to place a circuit voice call while accessing data networks via the Wi-Fi or WiMAX network).

In summary, the type of terminal used to support the convergence solution will also impact the necessary support from the network side. If the convergence application implies using a dual-radio terminal, then seamless mobility and a seamless user experience can be provided without the need for radio access networks to cooperate in preparing handovers or domain changes.

Summary

This chapter brought us one step further in our quest to understand the FMC technology components. We have analyzed the relevant aspects of cellular systems, compared circuit and packet technologies, and looked at radio access and mobility.

We then provided an in-depth discussion on the modern WiMAX and Wi-Fi technology landscape, paying special attention to voice service support. We looked at both the main VoWi-Fi enablers and the biggest technical roadblocks to its widespread deployment. Armed with this knowledge, we are now ready to proceed to the final destination of our journey through FMC technology, the *C* in FMC.

Chapter

5

The *C* in FMC

Thus they confronted: water and stone, prose and poetry, ice and flame differ less from each other.[1]

—Alexander Sergeevich Pushkin, Evgeni Onegin

The last step in our journey into FMC, after the review of its relevant fixed and mobile aspects in Chapters 3 and 4, respectively, is to provide a consolidated view on how fixed and mobile networks come together in the delivery of converged telecommunication services. This will include discussing the background of modern techniques enabling convergence and their application in real-life deployments.

We start with the discussion of the FMC systems based IMS and UMA/GAN. We then consider the converged solutions based on VoIP service delivered over multi-access fixed and mobile systems, including both Wi-Fi and cellular. Finally we will take a look at femtocell technologies, which lately gained prominence as a major FMC enabler.

The section that follows opens the chapter with an overview of convergence technology fundamentals.

Convergence Technology Fundamentals

The delivery of converged services implies the involvement of both the end-user device and the network in creating the *converged communication experience,* allowing the user, either transparently or with explicit actions (depending on each service definition), to access a service through a variety of access networks. Essentially service is always available wherever there is coverage via one of the *converged access technologies.* Such availability is achieved potentially via different levels of fixed and mobile access integration.

A converged service and applications also benefit from information stored in user profiles and some context information such as user location and real-time availability

---

[1] Они сошлись: вода и камень, Стихи и проза, лед и пламень. Не столь различны меж собой.

status (also known as presence—see Chapter 6). The combination of user context and user profile allows an operator (and the subscriber, if the user profile is user-customizable) to define a number of converged access policies and reachability profiles, depending on location, time of day, user preferences, and the access network type the user is camped on at a given moment.

Various technologies have been proposed in the industry to enable fixed-mobile convergence. Most of them are more oriented to converged services delivered by a mobile network operator (like UMA/GAN), since the level of integration there is higher and implies the use of the existing mobile network *standard interfaces* such as the GSM A interface and the GPRS G_b interface.

Other technologies—such as the IMS-based approaches—are more neutral to the type of operator delivering the service and can be adopted more widely across the industry, as the level of integration with the underlying fixed or mobile networks is looser, because the converged service experience is realized at the Service/Application layer.

Let's then delve into the analysis of service availability and details of specific levels of integration and their impact on the user experience (which we are going to explore in detail in the following chapter).

Levels of Integration

The success of a given FMC solution is a matter not only of making sure subscribers can access a service via particular access methods, but also of making network selection and switchover seamless and unnoticeable unless explicit notification is desired as part of the service definition ("you are now in the home zone where calling is free"). The seamless mobile experience users are accustomed to in a cellular environment should therefore be preserved and replicated as closely as possible in other access networks and application domains.

Of course, this broad requirement presents significant technical and usability challenges. Solutions to address them have been found and keep being identified as the industry successfully deals with the intricacies of delivering the converged network service experience.

From the integration point of view, convergence solutions can be broadly classified as "application-level" solutions (where an application on the terminal and application servers in the network cooperate in delivering the converged service experience), or "vertically integrated," that is, based on the user terminal cooperating with the network below the Application layer. In this model applications are not really required to support the delivery of the converged service experience, as the underlying infrastructure creates the appearance of a seamless connectivity via heterogeneous access networks. The former case is supported by technologies such as Voice Call Continuity (VCC), and the latter case is supported by UMA/GAN and femtocell approaches described later in the chapter.

Vertically integrated approaches can be then seen as ways to interface to the existing core infrastructure without the need for fundamental changes in network operation. This leaves the service delivery paradigm virtually unchanged and keeps it independent of the method of access. In other words, the investment necessary to bring vertically integrated solutions to life is concentrated in the access, terminals, and business models.

The application-level solutions, conversely, tend to leave the access network invariant to the deployment of converged services (no need of special access controllers or devices other than classic wireless routers with DSL or LAN interfaces in the premises where service is accessed through noncellular transport). These solutions, however, require terminals to have special software clients, and the core network to cooperate with these devices at the application level. The core network also must make sure that service delivery is uniform and service settings remain synchronized in the converged networks' domains.

VoIP-Based Convergence

The transition of traditional PSTN telephony to VoIP is contagious. The rise of interest in Voice over Wi-Fi is the latest evidence of this trend. The cellular industry is not far behind with a slew of new radio interfaces focused on the support of multimedia services such as CDMA EV-DO revA (also known as DOrA, to be used by the operators relying on 3GPP2 standards) and HSPA (to be adopted by 3GPP-bound operators) being rolled out around the world. The practical deployments of the VoIP-based solutions in cellular environments, however, have to deal with many issues, both technical and nontechnical, mostly absent in the fixed environment.

For one thing, cellular operators tend to maintain more stringent control over the access to the radio resources necessary to provide cellular services than wireline operators do over local loop or cable. The reason for that is quite simple, in that expensive licensed spectrum is shared among subscribers, and operators seek to maximize the return on their investment.

Second, the currently deployed cellular networks cannot fully support the VoIP service. For instance, they are lacking mechanisms to support guaranteed QoS for packet data services. On the other hand, although the new technologies capable of supporting VoIP in cellular environments[2] are already available, they are still months if not years away from widespread deployment, thus making cellular VoIP service support very limited, and only available in some vertical segments of the market, in the short term.

Finally, cellular VoIP requires new devices or new clients in the existing devices, which in turn calls for significant investment and long-term effort to be put in place by device manufacturers and their suppliers or independent software vendors.

For these reasons it can be argued that VoIP in a wireless environment, at least in the coming years, will be less likely to experience the level of freedom and growth we are witnessing today in wireline broadband. Despite that, the long-term future of cellular wireless VoIP is bright. We are convinced that competitive forces and economies of scale will eventually lead to parity between VoIP services over fixed and wireless networks and even the potential of uniform treatment of *non-operator-controlled* VoIP applications.[3]

---

[2] DOrA technology, for example, supports QoS standards necessary for differentiated VoIP traffic treatment.

[3] Imagine a commercial cellular VoIP offering from Skype.

Unlike non-operator-controlled VoIP, the *operator-controlled* VoIP services[4] require the deployment of specific telephony applications and the necessary interworking infrastructure needed to interface to the legacy circuit-switched voice subscribers both on cellular networks and in the PSTN. It is also necessary to ensure voice call continuity in areas with mixed VoIP and circuit coverage, or in other words, seamless handover of active calls between VoIP and legacy circuit TDM.

However, once IMS-based VoIP is uniformly deployed in the converged cellular and fixed networks, the resulting access independence of the converged core will ensure the right user experience. The mechanisms to ensure service continuity and active call handover between cellular and noncellular can at that point in time be implemented at the IP layer, using mature technologies such as Mobile IP (see RFC 3220 [22]) or other IP mobility support protocols.

Until that happens, the industry will have to deal with a difficult task of building FMC solutions combining dissimilar *packet* and *circuit* voice services implemented over heterogeneous access networks. The rest of this section is devoted to the discussion of such solutions.

The Dual Nature of Dual Mode

The contemporary voice FMC solutions supporting operation over Wi-Fi and cellular access networks are most often called *dual-mode* solutions. Such dual-mode FMC solutions can be based on a variety of technologies. The most prominent are the two standards-based approaches: One is built around IMS; the other is based on UMA/GAN, defined for the GSM cellular systems. Both approaches effectively converge circuit cellular voice and VoIP over Wi-Fi/broadband by "hiding" the Wi-Fi access media and signaling from the cellular core network.

However, that's where the similarity ends. While UMA/GAN attempts to tackle the problem of delivering converged services from the lower layers of the protocol stack, the IMS enables convergence at the higher layers. The IMS VoIP approach places the handling of the voice call in the IMS core network and allows for a significant part of the voice traffic to be offloaded from the cellular network core. Unlike IMS, the UMA/GAN standard enables the support of a Wi-Fi infrastructure by making it look like a set of GSM base station controllers, thus making Wi-Fi appear as just another 3GPP radio interface and requiring all traffic to traverse the 3GPP core network.

To better understand these aspects and more, in the following sections we discuss the fundamentals of both IMS and UMA/GAN FMC methods.

IMS and MMD Fundamentals

The IP Multimedia Subsystem (IMS) and the Multimedia Domain (MMD) are the 3GPP and 3GPP2 versions of the same thing, that is, an IP- and SIP-based system

---

[4] IMS is on target to become the de facto platform for delivery of cellular VoIP services (along with a variety of other multimedia services).

defined to handle multimedia signaling in the wireless domain. This system supports communications among SIP user agents accessing an IP network and various SIP servers within the network, using the 3GPP and 3GPP2 systems' packet data networks' access services.

Standardization   There are only minor differences between the 3GPP and 3GPP2 versions of IMS, and from almost every practical angle these versions can be considered similar, if not identical. There is in fact a cooperation agreement in place between 3GPP and 3GPP2 organizations, whereby 3GPP2 adopts the 3GPP IMS specifications, only with minor modifications to suit the unique 3GPP2 market needs.

Over time, the intrinsic access-independence of IMS (albeit IMS still has some access-specific details surfacing mostly at the Protocol layer) has prompted its adoption as a core enabler for other wireless access technologies such as WiMax and even for wireline networks.

It is important to point out that the proliferation and applicability of IMS to multiple access technologies has been its goal from the beginning. IMS was originally defined by 3GPP around the year 2000, following intense activity by the 3G.IP industry focus group, initially driven by AT&T, at that time interested in both wireless and cable networks. Therefore, the driver behind the standardization effort was to define a solution that, while delivering an IP-based platform for wireless multimedia services, as required by the wireless arm of AT&T, could also be used for other access technologies, especially cable.

It is no surprise, then, that as of late the wireline operators' community is increasingly considering IMS as the platform for their VoIP services. For instance, the BT 21st Century network, the first major-scale IP transformation project in the world carried out by an incumbent wireline operator, will migrate over time to IMS after an initial deployment based on an architecture closer to the server model (introduced in Chapter 3). Cable operators are also embracing IMS (as it was meant to be from day one, now we can say!), through their CableLabs industry forum.

But why are all these different branches of the telecom industry embracing the IMS and not its alternatives? There are two main reasons:

- The IMS is not only a standardized architecture for delivery of VoIP service, but also a general platform supporting all kinds of IP-based multimedia applications (text, images, instant messaging, conferencing, presence, and video telephony).

- The IMS core network can be shared by multiple access technologies, so it can be deployed by operators intending to pursue the convergence path, seeking to enter triple or quadruple plays, or just looking for more flexibility in the "last mile."

In the following sections we provide a technical overview of the IMS architecture, point out its differences from MMD, and discuss the IMS-based FMC solutions.

Architecture and Components   As evident from Figure 5.1, the IMS architecture is quite complex.

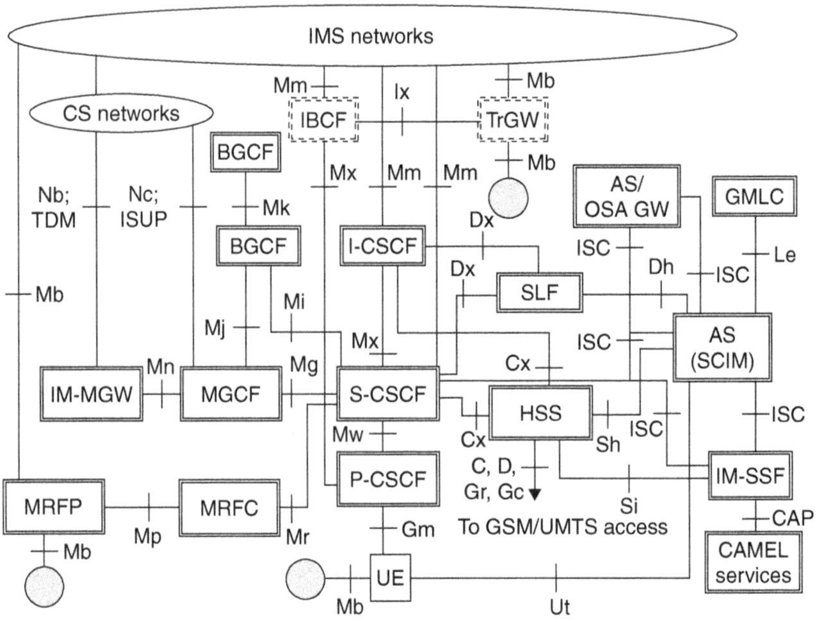

Figure 5.1 IMS architecture diagram

At the heart of IMS is a SIP server, called Call Session Control Function (CSCF). There are three instances of CSCF:

- Proxy CSCF (P-CSCF)
- Serving CSCF (S-CSCF)
- Interrogating CSCF (I-CSCF)

There are other important elements in the architecture, and these are described in the remainder of this section.

P-CSCF   The P-CSCF is the first contact point for the UE within the IMS. It acts as a SIP proxy (see RFC 3261 [23]), and it interacts with the admission control subsystem, so that only the media components authorized by the IMS are handled within IMS, with the appropriate level of QoS applied. This is enforced by a gateway the P-CSCF interacts with (e.g., the GGSN in 3GPP access systems). The UE therefore exchanges SIP signaling messages with the P-CSCF. The IP address of the P-CSCF is provided by the access network during the *terminal configuration* phase, when the UE gets access to IMS using the access network.

The P-CSCF may also behave as a user agent (see RFC 3261 [23]). In abnormal conditions it may in fact terminate and independently generate SIP transactions (e.g., to handle the case of a terminal which has lost radio coverage in a 3G system). The P-CSCF also is an endpoint of *security associations* between the IMS and the IMS-capable UE, to maintain SIP sessions' confidentiality over the access networks.

To enable NAT traversal, the P-CSCF may also act as a STUN server. The P-CSCF can be located in the *home network* or in the *visited network* if the UE IP address is assigned locally in the visited network (for example, based on a roaming agreement with the home operator). This roaming configuration, however, is almost never used in currently deployed 3GPP networks, where the home-based GGSN roaming is the defacto configuration, so roaming subscribers use their IP addresses as assigned by their respective home networks. When the P-CSCF is located in the visited network, it has to support charging detail records generation to allow a visited network operator to share revenue with the home operator.

S-CSCF   The S-CSCF role in IMS architecture is to enable services. It behaves as a *registrar* as defined in RFC 3261 [23]; i.e., it accepts registration requests and makes the registration information available via the location server (which is embodied in IMS by the Home Subscriber Server, or HSS described below), accessible via the C_x interface. It may perform barring functions, in that it can reject IMS communication to/from some well-known per-subscriber SIP user identities. An S-CSCF may also behave as a SIP Proxy server or a SIP UA (to independently initiate or terminate SIP sessions) and support interaction with service platforms via the SIP-based *ISC* (IMS Service Control) interface as depicted in Figure 5.1.

The S-CSCF can send SIP signaling for further processing to an *application server (AS),* e.g., based on static or dynamic rules (known as *initial filter criteria,* or IFC) provisioned on the S-CSCF, or based on per-user rules downloaded in the S-CSCF from the HSS when a SIP UA registers with the S-CSCF. The interface between the S-CSCF and an AS is known as an *ISC (IMS Service Control)* interface and is a SIP-based interface. Application servers use SIP signaling to implement a service logic. They interpret SIP messages from the S-CSCF and take specific actions (such as establishing third-party calls, invoking the usage of media resource functions, etc.).

The S-CSCF also performs call-routing functions, in that it can identify the entry point for the network of the SIP session destination (when the user does not belong to the same network as the source user of a SIP session handled by the S-CSCF). This may also be helped by a breakout gateway control function (BGCF) when the destination is a PSTN or a circuit-switched cellular network user. The S-CSCF may also be used for transit IMS scenarios and would act as a switch for transiting SIP sessions. The S-CSCF also generates call detail records (CDRs), which are the basis for retail charging for the IMS operator subscribers.

I-CSCF   Finally, to conclude the description of the various CSCFs, the I-CSCF is a SIP location function located at the edge of an IMS network. Even sessions originated by a roaming user who is using a local P-CSCF in the visited IMS network will be directed to the I-CSCF of the visited network to contact subscribers of the visited network (in fact, this roamer's SIP signaling is routed to its home network S-CSCF first, before interacting with the visited network entities, as IMS enforces home network control). An I-CSCF assigns a UE to an S-CSCF while a UE is attempting registration, and subsequently all SIP requests received by an I-CSCF from another network for that UE are routed toward this S-CSCF. This routing function is performed by obtaining the address of the S-CSCF from the HSS via the C_x interface. The I-CSCF also generates CDRs, which are useful for billing reconciliation and peering charges settlement between operators.

IBCF   To capture the role of session border controllers (SBCs), which are used (between operators or between the IMS operator and its subscribers), the IMS standards identify an additional function named the *Interconnection Border Control Function (IBCF)*. This function is collocated with (or represents) the entry point to the IMS network. The interfaces M_x toward the CSCFs and the BGCF are represented in the IMS architecture to underscore the logical separation from the CSCF. In physical implementations, however, both the BGCF and the CSCF are often supported by the same physical platform as the IBCF.

The IBCF provides functions necessary to perform interconnection between two operator domains, such as two IMS operators or an IMS operator and an ISP using SIP signaling to support VoIP applications. For instance, it can enable communication between IPv6 and IPv4 SIP endpoints, hide network topology by performing NAT functions such as IP address and (or) port translation, or act as a SIP firewall and generate CDRs related to peering between operators.

An IBCF interfaces to a *transition gateway (TrGW)* via the I_x interface. The TrGW handles media and is controlled by the IBCF. The TrGW also provides functions like network address/port translation and IPv4/IPv6 protocol translation. It may also implement and enforce firewalling rules (e.g., opening and closing pinholes on the media path, based on IBCF decisions). To date, since I_x is not yet specified down to the protocol level in the standards, all physical implementations of IBCF and TrGW are supported on the same physical platform, just as with BGCF and CSCF.

AS, SCIM, and IM-SSF   An application server is a SIP B2BUA that can process incoming SIP messages and, based on the information in a SIP message, generate new SIP messages and take actions based on the service logic it implements. It can also offer open interfaces toward third-party applications in the form of an API (application programming interface). In this sense it can act as a broker toward other servers. The importance of the AS function in FMC will be clarified in the section "VCC."

An S-CSCF may interact via the *ISC* interface with multiple application servers directly, or via an application server acting as a broker. An application server acting as a broker is considered, in standards parlance, to implement the Service Capability Interaction Manager (SCIM) feature. An AS implementing the SCIM feature can interact with a variety of application servers over ISC interfaces to deliver the desired service.

In the same way as an S-CSCF, a SCIM may interact with SIP application servers, Open Service Architecture (OSA) gateways, or IM-SSFs (IMS Service Switching Functions). The IM-SSF offers an ISC interface toward an S-CSCF or a SCIM, and a CAMEL Application Part (CAP, an SS7–based Intelligent Network signaling) interface toward the CAMEL Service Environment.[5] The IM-SSF downloads from the

---

[5] CAMEL (standing for Customized Applications for Mobile Network Enhanced Logic) services are Intelligent Network services in a 3GPP mobile environment. Their equivalents in 3GPP2-based systems are governed by WIN2 standards.

HSS the CAMEL triggers for a specific user and then arms them via the S_i interface. Similarly, an AS can retrieve application data for a subscriber from the HSS via the S_h interface and access user location information available in the Gateway Mobile Location Center (GMLC) via the L_e interface.

MRFC and MRFP   While conferencing capabilities can be implemented in IMS by means of a dedicated AS, with the terminal setting up the conference via the U_t interface with the application server, there is a function in the IMS that is specifically defined for that very purpose, the Multimedia Resources Function Controller (MRFC) combined with the Multimedia Resources Function Processor (MRFP). The MRFC and MRFP can also be used to deliver tones and announcements and to generate multimedia content necessary for a specific IMS session. Unfortunately, the specification of the M_p interface between the MRFP and MRFC has been developing quite slowly, and to date it is still a bit undefined, so in most cases the market has gone in the direction of proprietary solutions based on application servers hosting media-processing capabilities.

MGW and IM-MGW   In IMS, the interworking with CS networks, whether toward classic GSM and CDMA networks and the PSTN based on TDM and ISUP, or a BICC-based bearer-independent, circuit-switched core network (based on the Rel-4 specifications of 3GPP, namely 3GPP TS 23.205 [184]), is made possible by the Media Gateway Control Function (MGCF) and the IMS Media Gateway (IM-MGW). Many sources also refer to MGCF as a *softswitch,* as it performs functions equivalent to that of a class 5 TDM switch in the PSTN.[6]

Media gateways are controlled by MGCF and perform transcoding between different speech formats used in the CS and IMS networks. They may be put in the path also when different IMS networks adopt different voice encoding formats (e.g., a 3GPP2 network may use EVRC [65] encoding for speech while a 3GPP network may use AMR [63] or WB-AMR [64]).

Since IMS-based FMC scenarios in the immediate future involve interworking between CS networks and IMS-based VoIP, the media gateways play a fundamental role in FMC.

HSS and SLF   The HSS—also referred to as User Profile Server Function, or UPSF, by TISPAN—is a subscriber database providing subscriber data to the IMS elements that require this information while handling calls and sessions. In addition, the HSS performs authentication and authorization functions and can act as a SIP location server.

The HSS was originally conceived as an evolution of the traditional HLR and AUC, and as a consequence, it is also capable of supporting the legacy MAP signaling interfaces toward MSCs (*C* and *D* interfaces, where *D* is used toward gateway MSCs), SGSN (via the G_r interface), and the GGSN (the G_c interface) in GPRS.

---

[6] Softswitches and media gateways create their own complex field. If you are willing to probe further, you can refer to an excellent book by Frank Ohrtman, *Softswitch: Architecture for VoIP* (McGraw-Hill, 2002).

It should be noted that the Home Subscriber Server (HSS) in Figure 5.1 is also accompanied by a Subscriber Locator Function (SLF). An SLF is used in IMS deployments with more than one addressable HSS. In this case, network entities that require access to HSS subscriber data need to access the SLF to identify the HSS in which a specific subscriber's data is actually stored. The D_x interface is used to access the SLF from I and S CSCFs. The D_h interface is used to access the SLF from an AS.

Interface Summary   Table 5.1 provides a synopsis of the interfaces discussed in this chapter. There are also other interfaces, but for these we invite the interested

TABLE 5.1   IMS Interfaces and Protocols

Interface	Description	Protocol
C_x	Reference Point between a CSCF and an HSS	DIAMETER
D_x	Reference Point between an I-CSCF and an SLF	DIAMETER
G_m	Reference Point between a UE and a P-CSCF	SIP
ISC	Reference Point between a CSCF and an application server	SIP
I_x	Reference Point between IBCF and TrGW	Not specified in 3GPP Rel-7
L_e	Reference Point between an AS and a GMLC	Mobile Location Protocol (MLP version) carried over HTTP
M_a	Reference Point between an AS and an I-CSCF	SIP
M_b	Reference Point to indicate IP packets carrying media	IP
M_g	Reference Point between an MGCF and a CSCF	SIP
M_i	Reference Point between a CSCF and a BGCF	SIP
M_j	Reference Point between a BGCF and an MGCF	SIP
M_k	Reference Point between a BGCF/IMS ALG and another BGCF	SIP
M_m	Reference Point between a CSCF/BGCF/IMS ALG and an IP multimedia network	SIP
M_n	Reference Point between an MGCF and the IM-MGW (Mc is the interface between MGCF and MGW in server model or Rel 4 bearer-independent core network.)	H.248
M_p	Reference Point between an MRFC and an MRFP	H.248
M_r	Reference Point between a CSCF and an MRFC	SIP
M_w	Reference Point between a CSCF and another CSCF	SIP
M_x	Reference Point between a CSCF/BGCF and an IBCF	SIP
S_h	Reference Point between an AS (SIP AS or OSA CSCF) and an HSS	DIAMETER
S_i	Reference Point between an IM-SSF and an HSS	DIAMETER
U_t	Reference Point between a UE and an application server	Application dependent (e.g., carried over HTTP)

readers to probe further in the IMS literature, such as, but not limited to, 3GPP TS 23.228 [66], 3GPP TS 24.228 [67], and books such as [68] and [69].

Subscriber Identity   Within an AS, a CSCF, an HSS, or a UE a particular user is uniquely identified by means of a *user identity*. A user identity can be in the format of a SIP URI–RFC 3261 [23] or a Tel URI–RFC3966 [30] defined in Chapter 3. A SIP URI has the following format:

```
SIP URI = sip:x@y:Portnumber
Where x=Username and y=hostname|domain
```

Examples of SIP URIs are

```
sip:john.doe@212.123.1.213:1218
sip:helpdesk@bigphoneco.com
sip: sip:234150999999999@ims.mnc015.mcc234.3gppnetwork.org
```

IMS supports both private and public user identities. Public identities are advertised externally to the IMS, are in the SIP URI format, and are public to other users, much in the same way as e-mail addresses and telephone numbers are public. Private user identities take the format of an NAI–RFC4282 [70], i.e., "username@realm." They are associated with the user subscription and used by the UE to register with the network. They are also used for authentication of the user and to access the HSS data for the subscriber.

The private user identity is stored in the I-SIM application (IMS – SIM), provided with the UICC (User Identity Chip Card—a smart card provided to 3GPP and some 3GPP2 operators' customers, commonly known as SIM—Subscriber Identity Module—by most GSM networks users) given to the user as part of the IMS service subscription. Users may be contacted through a variety of *public* user identities that they give out to other users. These identities can be associated with the same service or with different IMS services. At least one public identity needs to be stored on the I-SIM application. The network then translates these to a single private user identity for its internal operation (e.g., charging, database access, etc.). Note that *private* user identities are not used in SIP messages.

Some IMS services also require the capability to interact with a *group* of users that is dynamically or statically defined. An example is a chat list, where a number of users can post messages to the list and receive messages from the list. This is also necessary for multimedia conferencing applications and in general for group services. For this purpose, the concept of *Public Service Identity (PSI)* has been defined in IMS. A PSI is in the form of a SIP URI; it could be, for instance:

```
chatlist1@bigFMCoperator.com
```

Messages addressed to a PSI apply to a group of users or a particular service hosted on a SIP AS. The IMS can support addressing in SIP messages using Public User Identities or Public Service Identities.

Access-Specific Procedures   While IMS is generally defined as an access-independent architecture, it was necessary to identify some access-specific procedures needed to cover access-specific security mechanisms, access-specific information, and P-CSCF discovery. The 3GPP IMS specifications also define the use of a *P-Access-Network-Info* header, part of the Private Header (P-Header) Extensions to the SIP for 3GPP (RFC 3455 [71]).

This access network–specific information element can identify the cell ID the user is in, the specific technology being used, and other access-specific parameters. The information about the type of access technology can be utilized by the network to customize service operation to the capabilities of the access network. The information to be inserted by a UE in the P-Access-Network-Info is specified in 3GPP TS 24.229 [72]. The syntax of this header, as described in 3GPP TS 24.229, is the following:

```
access-type = "IEEE-802.11" / "IEEE-802.11a" / "IEEE-802.11b" / "IEEE-802.11g" /
"3GPP-GERAN" / "3GPP-UTRAN-FDD" / "3GPP-UTRAN-TDD" / "ADSL" / "ADSL2" /
"ADSL2+" / "RADSL" / "SDSL" / "HDSL" / "HDSL2" / "G.SHDSL" / "VDSL" / "IDSL" /
"3GPP2-1X" / "3GPP2-1X-HRPD" / "DOCSIS" / token
access-info = cgi-3gpp / utran-cell-id-3gpp / dsl-location / i-wlan-node-id /
ci-3gpp2 / extension-access-info
extension-access-info  = gen-value
cgi-3gpp = "cgi-3gpp" EQUAL (token / quoted-string)
utran-cell-id-3gpp = "utran-cell-id-3gpp" EQUAL (token / quoted-string)
i-wlan-node-id = "i-wlan-node-id" EQUAL (token / quoted-string)
dsl-location = "dsl-location" EQUAL (token / quoted-string)
ci-3gpp2 = "ci-3gpp2" EQUAL (token / quoted-string)
```

This header specifies the access technology type using the access-type parameter, and some access-specific information using the access-info parameter. The parameters can take any of the alternate values described already. The "gen-value" parameter can be an arbitrary defined value, which inevitably can only be used intradomain or within a federation of operators who agree on its meaning.

IMS vs. MMD   As we discussed at the beginning of this section, the IMS has been initially defined within wireless standards bodies defining cellular systems. Since today there are two main standards bodies for cellular systems specifications definition, 3GPP and 3GPP2, there are consequentially two flavors of IMS, which are almost identical with the exception of specific deployment-related aspects. These differences, however, do not affect interoperability between the 3GPP and 3GPP2 versions of IMS.

Of course, one such difference is in the name: IMS in 3GPP and MMD in 3GPP2; but certainly this has nothing technical to it! Other differences are more substantial. For instance, 3GPP IMS mandates IPsec to be used between the UE and P-CSCF to secure IMS signaling, while 3GPP2 allows for P-CSCF and UE to negotiate other security mechanisms using RFC 3329 [73] (an IETF recommendation on how to set up security mechanisms for SIP). Also, 3GPP IMS terminals have UICCs, whereas MMD does not require a smart card, as 3GPP2 UEs may not be equipped with a Removable User Identity Module (R-UIM). So, as a consequence, in MMD, subscriber identity information can be stored in the terminal itself or in the R-UIM; albeit 3GPP2 supports UICC + I-SIM for operators that choose that method.

In 3GPP, it is also possible for a UE to access the IMS without an I-SIM. For this purpose, the 3GPP IMS creates temporary Public/Private IDs to support terminals without

I-SIM applications. This was not a feature of MMD Rev 0 (the first release of MMD), but it is now supported in MMD Rev A. However, the way MMD generates these identities is different from 3GPP IMS because 3GPP and 3GPP2 accesses use very different ways to identify a subscriber. 3GPP systems use ITU-T recommendation E.212 [74]–based International Mobile Subscriber Identity (IMSI), while 3GPP2 systems use a Mobile Identity Number (MIN)–based Mobile Subscriber Identifier (MSID). It should be noted that 3GPP2 specifications support IMSI-based operations based on IS-751 [75], but the adoption of this has been quite limited.

The HSS in 3GPP also offers legacy interfaces, as it behaves as a classic HLR toward the CS and PS domains of the 3GPP access system. In 3GPP2, however, the HSS acts as a AAA server and subscriber database only for IMS and the 3GPP2 PS domain, thus representing a totally new function in the network, dedicated to PS services.

The UE in 3GPP2 also supports some 3GPP2-specific parameters in the SIP P-Access Network-Info header information provided in registration messages to the P-CSCF. There are also differences in the P-CSCF discovery procedures, in that 3GPP2 supports static configuration and DHCP, while 3GPP can use the 3GPP access signaling to discover the P-CSCF (the P-CSCF IP address can be provided in the signaling involved in PDP context activation).

Finally, in 3GPP, the P-CSCF can be in the visited network or in the home network, while in 3GPP2, the P-CSCF can only be in the home network when Mobile IP is used, and in the visited network when Simple IP is used. So, in 3GPP2, the roaming model includes a PCC interface between the visited network and the home network to enable the enforcement of PCC for roaming users using the Simple IP service.

Table 5.2 summarizes the differences between IMS and MMD architectures.

IMS over Wi-Fi and its deployment aspects   The deployment of VoIP in the cellular environment will not happen, at least initially, in the "uncontrolled" way we are experiencing today with fixed VoIP offerings and VoIP over Wi-Fi access in the home zone and hotspots. Cellular carriers are expected to tightly control it using the IMS as a platform.

TABLE 5.2   IMS vs. MMD—Main Differences

Functionality	IMS	MMD
SIP signaling security	IPsec used between the UE and P-CSCF	IPsec used between the UE and P-CSCF; P-CSCF and UE can negotiate other security mechanisms using RFC 3329
Removable card for subscriber data	Yes (UICC + ISIM)	Sometimes not available
Access IMS without an I-SIM	Yes	Not possible in Rev 0 Possible in Rev A
HSS offers legacy interfaces (acts as HLR)	Yes	No
P-CSCF location	Location of the GGSN used to access IMS	Only in the home network when Mobile IP is used; in visited network for Simple IP

Rolling out VoIP using IMS (whether in a wireline, Wi-Fi, or cellular environment) requires the deployment of some specific telephony application servers and other elements of the interworking infrastructure needed to interface to both cellular and PSTN circuit-switched voice subscribers. This per se would not allow for a converged service experience, as the various types of access would not be blended transparently to the user of a dual mode Wi-Fi and cellular access-capable UE.

To ensure successful fixed mobile convergence while delivering voice services in a mixed cellular and Wi-Fi environment to users of dual mode UE's, it is therefore necessary to create a mechanism to ensure continuity of voice calls across technologies. Over cellular access it may in fact be possible that IMS-based voice service is not yet supported, while it is over Wi-Fi, when Wi-Fi is used to access the IMS over, for instance, a DSL access. As an aside, this same mechanism would also be required within a cellular access technology, between areas where the network has been upgraded to support VoIP and areas where service has to fall back to circuit-switched telephony, as no upgrade has taken place.

In addition, operator-controlled VoIP support in cellular or Wi-Fi environments requires the operators to comply with regulatory requirements for the provision of emergency services over packet accesses.

Focusing on Wi-Fi access, there exist three main deployment scenarios:

- Enterprise
- Hotspot
- Residential broadband or *home zone*

In an enterprise environment, the deployment of VoIP and FMC is normally not based on a commercial service unless outsourced to a third party, such as a VoIP operator or integrator, which may also offer other voice and data services to the organization, and comes under the CIO jurisdiction. A potential commercial service in an enterprise environment using the Wi-Fi access could be a wireless IP Centrex, whereby the PBX service is virtually hosted in the operator's network combined with in-house network of interconnected Wi-Fi access points.

In hotspot areas, the Wi-Fi operator could offer IMS FMC service, accomplishing two major goals. On the retail side, the offer of VoIP can provide access to voice services at a discount versus cellular access (and perhaps allow for very advantageous propositions for users frequently roaming abroad, when hotspots are operated internationally or when federations of hotspot providers come together to offer VoIP services also for roaming users). On the wholesale side, the hotspot provider may offer cellular network offload services, for those subscribers using dual-mode cellular/Wi-Fi terminals. This may also take advantage of Voice Call Continuity (VCC)[7] capabilities offered by the home cellular network.[8]

In a residential environment, VoWi-Fi is normally used as a cordless-like alternative to fixed VoIP service, when a wireless DSL router or cable modem is deployed, or as part of a converged solution offered and controlled by a cellular network operator,

---

[7] The voice call continuity feature is described later in the chapter.

[8] Chapter 6 provides further discussion on business aspects of VoIP deployment and convergence.

like IMS-based or GAN/UMA solutions. To exemplify the usage of the P-Access net-work when using Wi-Fi access networks, the P-Access-Network–Header used in a Wi-Fi/WLAN environment is normally set to

```
access-type = "IEEE-802.11" / "IEEE-802.11a" / "IEEE-802.11b" / "IEEE-802.11g"
access-info = i-wlan-node-id
i-wlan-node-id = "i-wlan-node-id"
```

It should be noted that in the case where Wi-Fi is simply a connectivity medium to a Wi-Fi-capable DSL router or cable modem, the P-Access-Network-Info will be set to a value that reflects the use of DSL or cable to access the IMS. So, from a system perspective, the fact that a subscriber is using Wi-Fi in some residential environment as a way to connect user devices to the fixed access is transparent to the IMS.

These observations point to the fact that the handling of Wi-Fi as a distinct access technology will normally apply when the Wi-Fi access is offered directly to customers as a service by itself, as in the hotspot Wi-Fi service model. It should also be noted that 3GPP has defined a specification for the interworking of these commercial WLAN access networks with 3GPP systems' packet core networks. These interworked WLAN access networks are commonly identified as I-WLAN (defined in 3GPP TS 23.234 [76]). I-WLAN is therefore the way a 3GPP IMS system would classify the access to a 3GPP IMS via 3GPP TS 23.234 compliant Wi-Fi access.

IMS-Based Convergence

As we established in the preceding section, the IMS offers a unique capability to act as a single service delivery environment to IP endpoints. These endpoints could be client devices accessing the IMS core via IP-based access networks or termination points of SIP signaling with RTP media streams at a *media gateway* for interworking with the PSTN or cellular core networks. Since the IMS can be used to provide service to both packet domain users (IP endpoints) and circuit-switched users via media gateways, it offers certain intrinsic capabilities to converge the circuit- and packet-based services. It is therefore no surprise that the IMS is increasingly used as a foundation of converged networks.

A well-implemented FMC solution should be transparent to the end user, and the resulting converged communication experience should appear *uniform* with no need for the subscriber to be *aware* that two or more different domains are being used when placing a phone call or starting a data application. Switching between different access technologies or between circuit and packet domains should be seamless even during an active call handover (e.g., because of the change in coverage, or due to certain conditional switchover triggers such as time of day or changing location to a home zone).

Today's typical IMS and SIP VoIP FMC solutions are based on the combination of VoWi-Fi access and IMS core with cellular infrastructure tied together by means of a certain *Convergence Gateway* function and an FMC application server, anchoring calls across the domains (see Figure 5.2). Such an AS and gateway serve as a hub for both SS7 and SIP signaling, allowing for call establishment between different types of terminals and handling of services and calls placed to VoWi-Fi clients through the cellular infrastructure.

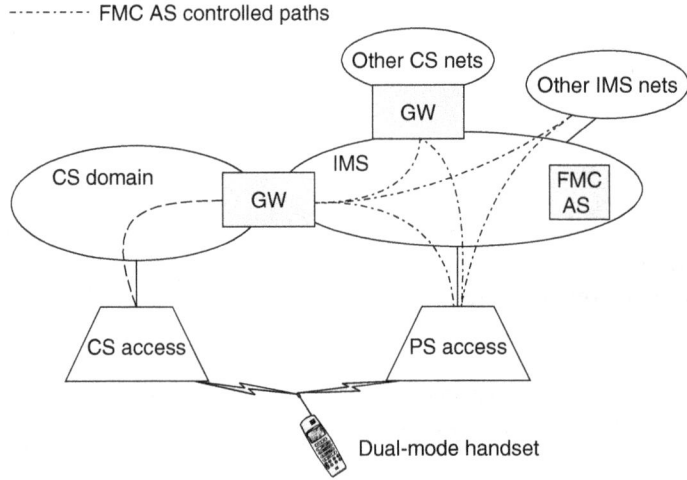

Figure 5.2   Typical SIP-based FMC architecture

Unlike in UMA/GAN solutions, the bulk of VoIP media and signaling traffic in IMS-based FMC solutions is routed directly (i.e., without transitioning through a cellular core network) to PSTN gateways or other VoIP networks when the subscribers are connecting through noncellular access mechanisms. To access thus-defined FMC services, the subscribers must use dual-mode Wi-Fi/cellular devices.

While in recent years the industry came up with a variety of proprietary solutions based on the convergence gateway concept, a standard-based approach commonly known as Voice Call Continuity (VCC) has been recently defined by 3GPP. This approach is described in the next section.

VCC   VCC assumes the IMS as the centralized point of control for voice calls, both circuit-switched voice and VoIP. This control is implemented by deploying a VCC application server (VCC AS) in the IMS network. To make use of VCC, the subscriber needs a dual- or multimode handset capable of supporting VoIP over a packet access network (like Wi-Fi, WiMax, or the PS domain of a cellular access system), and circuit voice over GSM, UMTS, or CDMA systems. This terminal also needs a client capable to "glue" the circuit and packet voice streams together by interacting with the VCC application server in the core network as defined in VCC specifications 3GPP TS 23.206 [76] and 3GPP TS 24.206 [77].

The users of multimode handsets also need to obtain a VCC subscription with their service providers, so that their calls are anchored at the VCC application server. The VCC application can be invoked to establish and tear down call legs over the CS access (by controlling media gateways) or over the PS access, and update remote endpoints participating in the call with the VCC subscriber. A change of domain is implemented by, e.g., setting up a call leg on the IMS side, informing the remote endpoint that the call is now fully in IMS, and tearing down the leg on the CS side.

Voice calls can be both originated and terminated by a VCC subscriber camped on one or both of the CS networks or the IMS at any time. Therefore special procedures are needed to anchor the calls originated from the VCC subscriber and to terminate calls destined to a VCC subscriber. Call termination implies selecting onto which domain to try and deliver the call first, so a *domain selection* process is required while terminating calls. Call origination also requires the VCC UE to determine onto which domain to place the call (if more than one is available). In both termination and origination, operator policies and user preferences come into play and become part of the VCC service definition offered by the operator.

VCC Call Setup   The call anchoring process for VCC calls, depicted in Figure 5.3 for the case of a UE originated call, is quite complex and is worth discussing in detail.

To implement VCC call anchoring, it is necessary to have the IN implemented in the visited MSC. It is further necessary to ensure that the visited MSC supports certain trigger detection points (CAMEL trigger detection points in 3GPP systems and WIN2 trigger detection points in 3GPP2). When the VCC UE initiates a voice call using a circuit-switched technology, the appropriate IN service is triggered, which will cause

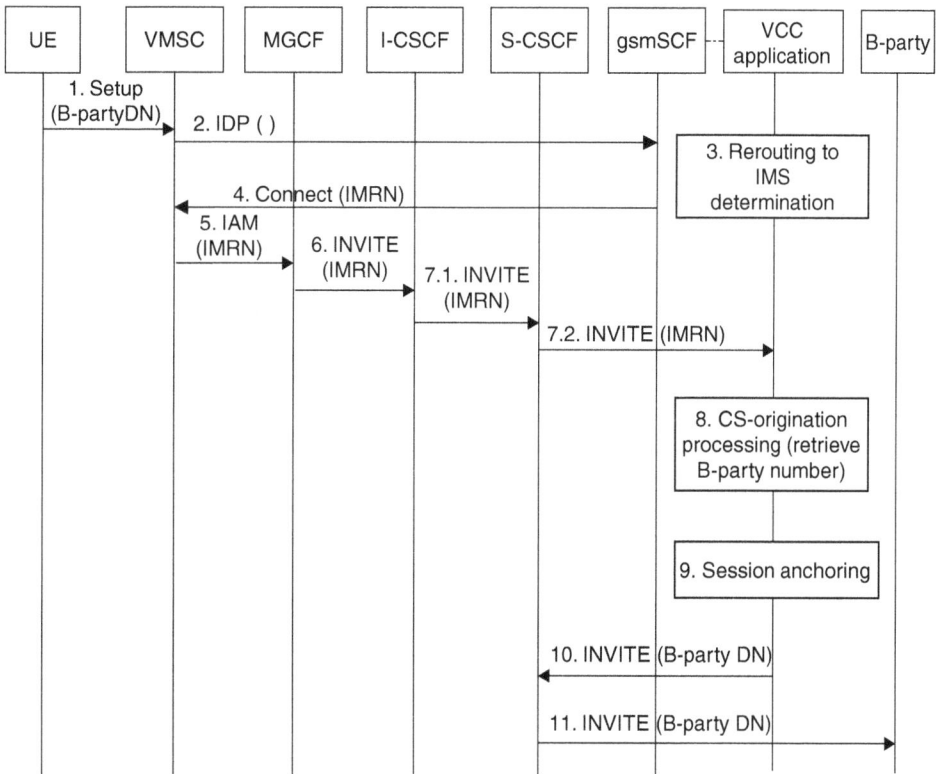

Figure 5.3   VCC call setup diagram

the MSC to query the gsm-SCF (in 3GPP systems, or the equivalent IN Service Control Point in 3GPP2 systems) to obtain the address to route the call to an MGCF, for anchoring and subsequent handling by the IMS. The address used to route the call to the correct MGCF provided by the gsm-SCF is called the IMS routing number (IMRN). This number is conceptually similar to the MSRN used in GSM to route incoming calls to the correct serving MSC for a mobile subscriber.

The MGCF then starts a conventional SIP call setup over the IMS core by sending a SIP INVITE toward an I-CSCF using the IMRN as the called party address. The I-CSCF or the S-CSCF (based on *initial filter criteria*) then invokes the VCC application by forwarding this SIP INVITE to it. Upon reception of the INVITE, the VCC application concludes the anchoring by forwarding the SIP invite back to the S-CSCF with the address of the called party instead of the IMRN. The S-CSCF forwards this SIP message in the direction of the called party.

This approach requires the allocation of the IMRN to the VCC subscriber at call setup, and a mapping of IMRN to the called party number to be available at the VCC application server. At the end of this transaction, IMS stores a state corresponding to this call.

It should be noted that the called party for such a call may be a PSTN or circuit cellular handset, so another MGCF needs to be involved in the call on the called party side or perhaps in some VoIP peering location. For calls initiated in IMS, the anchoring is simple and implies involving the VCC application in the handling of SIP sessions initiated by dual-mode UE. Note that not all SIP sessions traversing a CSCF and originated by a SIP endpoint are anchored at the VCC application, only those that are related to a voice call, which S-CSCF can discover by analyzing the SDP carried in SIP invite messages.

Once the call state is anchored in the VCC application, its *Domain Transfer Function* (see the next section) begins acting as a SIP B2BUA and assumes control of the *access leg* of the call (toward the VCC UE) and of the *remote leg* of the call (toward the peer of the VCC UE in the voice call). Voice calls bound for VCC subscribers in IMS can also be anchored following similar procedures.[9]

Domain Selection   When a call is inbound for a VCC subscriber, the VCC application must select whether to try to reach the UE via a CS or PS technology. The act of making this decision is known as *domain selection* in 3GPP jargon. The incoming calls themselves can be originated either from another IMS network or from a CS network. In the latter case an MGCF in IMS core would always be involved in the call path toward the VCC subscriber. The specific way a gateway MSC (GMSC) routes the call toward the MGCF and the criteria that could drive the selection are not mandated in the standards. For example, it could be based on assigning blocks of numbers to VCC subscribers.

The selected domain may not be the one the UE is currently camped on, as in fact, the decision is taken in the home network. Nor is it possible to accurately keep track

---

[9] Interested readers are invited to explore the details available in the 3GPP specifications [76] [77].

of location and access technology attachment status and reachability. If the attempt to contact the VCC UE fails in the selected domain, the UE is contacted in the other domain. If failure occurs in both domains, the call attempt fails.

VCC Domain Transfer   The *active call handover* between different access networks, known as *domain transfer* in VCC terminology, is one of the important aspects of 3GPP-defined FMC solutions and therefore should be considered here in detail. Active call handover or *seamless handoff* is defined as the ability to maintain a user's active calls or data sessions when the user changes to a different type of access network or migrates between different network domains.

When an active call is anchored at the VCC AS, it is possible to hand over the call to a different access network or perform domain transfer by using the method defined in 3GPP TS 23.206 [76]. If the dual-mode UE is in a PS voice call initiated over IMS and needs to transfer the call to CS (for example, based on certain preset policies or change in coverage), the VCC terminal attaches to the CS domain and initiates a CS call by sending a 3GPP 24.008 [180] SETUP message including a VCC Domain Transfer Number (VDN).

The CAMEL trigger at the MSC (armed for VCC subscribers) invokes the CAMEL service environment to provide a translation between the VDN and the Public Service Identity (PSI) of the VCC application Domain Transfer Function (DTF). The ISUP IAM message toward an MGCF used to access the home IMS environment uses the VCC DTF PSI as the called party. This message is then routed by the Home I-CSCF (or the S-CSCF, depending on implementations) toward the VCC AS. The VCC Domain Transfer Function of the VCC AS performs two actions, as shown in Figure 5.4:

- It updates the access leg by informing the remote party in the call (knowing that the call was anchored at the DTF) of the fact that the SIP session is now handled via the MGCF.

- It releases the source leg (clears the leg between the SIP UA in the VCC UE over the PS domain connection and the S-CSCF). As a result of this procedure, the call is now transferred on the CS domain.

Similarly, if the handset is in an active call on the CS domain, the transfer procedure from the CS domain to IMS (see Figure 5.5) starts with the VCC UE registering with the IMS core (if it was not yet registered) and then sending over the PS access network a SIP INVITE, including the VCC domain transfer URI (VDI), toward the S-CSCF (the VDI is also a VCC DTF PSI). The S-CSCF then invokes the VCC Application Domain Transfer Function, which performs the following two actions:

- It updates the access leg by informing the remote party that the call is now handled by the SIP UA in the VCC UE.

- It releases the source leg toward the MGCF.

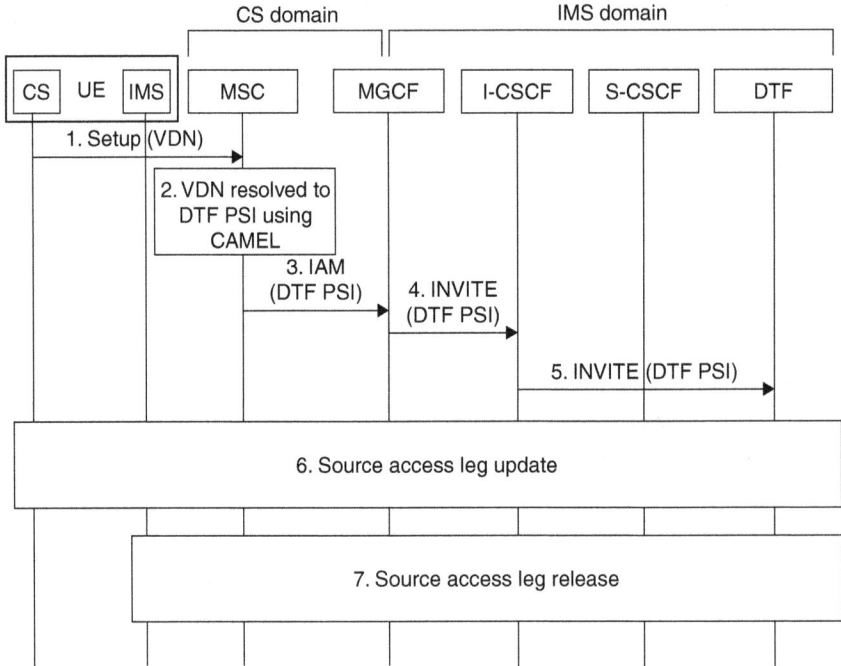

Figure 5.4   Domain transfer from IMS to CS

Both the VDI and the VDN are provisioned on the UE. These can be statically provisioned or dynamically assigned according to operator preferences (not defined by VCC specifications).

In order to realize the handover as a *"make before break"* process (that is, update the call access leg before the source leg is torn down), VCC, as defined in 3GPP Rel-7 in

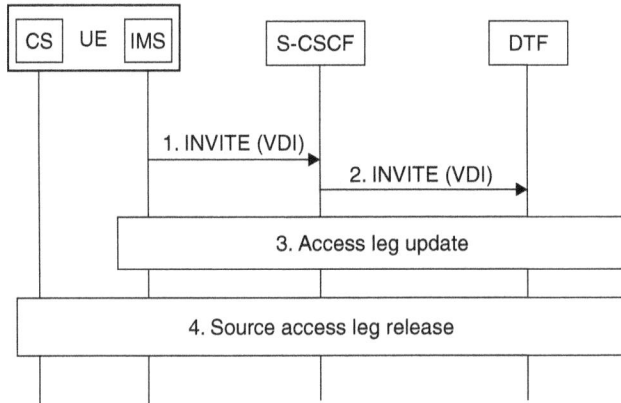

Figure 5.5   Domain transfer from CS to IMS

3GPP TS 23.206 [76] and 3GPP TS 24.206 [77], assumes that dual-mode terminals are capable of supporting concurrent access to the CS and PS technologies. If a sudden loss of coverage occurs in the access network supporting an active call, or if the terminal can only be active on one technology at a time, then the domain transfer will be based on *"break before make,"* and therefore will be less seamless.

Both the "make before break" and the "break before make" approaches assume that the alternative access network is available and therefore it is possible to hand over the active call by performing domain transfer. Needless to say, in the "break before make" case there could be perceptible interruption in the voice conversation.

VCC Applications Beyond FMC   In addition to converging cellular and VoWi-Fi telephony, VCC can also be used to transfer calls between the CS and PS cellular domains when an operator decides to migrate parts of its network to cellular VoIP using technologies such as UMTS HSPA, CDMA 1X EV-DO rev A (also known also as DOrA, or HRPD—high-rate packet data—specified in TIA IS-856 [78]), or LTE.

A UMTS terminal can maintain a CS call and the PS domain bearer needed to access the IMS concurrently active, thus meeting the applicability conditions of VCC Rel-7. A different discussion applies for the 3GPP2 VoIP deployment over DorA.[10] Since DOrA is a PS-only system, not unlike Wi-Fi networks, Voice over IP is the only way to support telephony service in this technology. Dual-mode DOrA/CDMA 1x terminals can only have one of the CDMA 1x CS or DOrA PS bearers active at a time. Therefore, in this case, the 3GPP Rel-7 VCC specification cannot be directly applied in 3GPP2 environments.

The VCC operation for 3GPP2 1x/HRPD dual-mode terminals (as well as for 1x/Wi-Fi) is therefore different than 3GPP VCC and is defined in a separate document [79]. The problem of supporting continuity of a voice call with dual-mode terminals supporting a single transmitter/receiver (also known as "Single Radio" dual-mode terminals, not allowing the simultaneous transmission/reception over the two technologies) is a general problem that the industry will need to face in the course of transition toward PS-only cellular systems. So it can be expected that a generic, standard way to support single-radio, dual-mode terminals will be defined.

The support of VoIP in cellular environments is a broad new direction for the industry as a whole, helping it to converge toward a single IP-based network to provide all services. That's why VCC received so much attention during its standardization process as a critical technology, not only enabling the short- and long-term objectives of FMC, but also providing building blocks for the technologies that follow on its footsteps.

Using elements such as VCC as a means to overcome the technical challenges of rolling out VoIP in a cellular environment, in a strict sense, does not represent an FMC application. Nevertheless, in our opinion, VCC has far-reaching impacts in its important role as a stepping stone in the evolution to fully converged seamless communications over a single IP-based network.

---

[10] Note that many aspects of CS and PS call interaction in CDMA DOrevA are still not fully defined in 3GPP2 at the time of the book writing.

Practical Deployment Considerations   Fixed operators that do not own and operate a cellular network, but still wish to provide their customers cellular services seamlessly converged with their fixed offerings, in most cases must rely on mobile virtual network operator (MVNO) relationships with a mobile operator. Their goal is to offer mobile services without owning the cellular access network and without giving up the ownership of subscribers to the cellular carriers. This can be achieved by establishing a wholesale agreement with a cellular operator, allowing the fixed operator to retain subscriber control. The FMC services enabled by such a relationship must therefore permit fixed operators to control HSS/HLR and back-end systems, allowing them to "own":

- All aspects of wireless account management and provisioning
- The ability to assign cellular subscriber numbers and identities
- Subscriber authorization and authentication
- Wireless applications and content hosting

The IMS and VCC architectures provide sufficient flexibility to satisfy these requirements. In the case of MVNO, mobile networks, as seen by the FMC fixed-line operator, become simply roaming partners.

The issue of synchronization or "centralized service control" is a very important one when the IMS-based VCC FMC solution is deployed, as the CS domain's supplementary services are working separately from the IMS-based simulation of the same services when the UE is on the IMS side, unless they are synchronized in a proprietary manner, or the service control is fully provided by the IMS. The latter option is still being standardized in 3GPP, and it represents a very important building block for the delivery of an FMC service fully compliant with the IMS framework and providing enough flexibility for fixed and mobile operators to enter in creative partnerships.

In contrast to IMS, solutions such as UMA/GAN do not allow a wireline carrier to exert control over cellular subscribers, essentially keeping the cellular operator in charge at all times. Furthermore, the UMA/GAN solution only addresses the GSM operator's markets, effectively making IMS VCC the only standards-based technology for FMC application by 3GPP2 operators. Having said that, we must recognize that such a narrow focus of the UMA/GAN standard made it extremely attractive for a specific target segment, that is, GSM operators, and caused it to become the foundation for the first commercially successful FMC deployments.

The following section considers the UMA/GAN technology in detail.

UMA/GAN

The idea of connecting to the GSM/GPRS core network via wireless access technologies using an unlicensed spectrum, such as Bluetooth or Wi-Fi combined with broadband, to deliver the same services as in the licensed spectrum always looked quite attractive to cellular network operators. These operators were willing to combine or extend their traditional service offerings over GSM cellular access with the inexpensive coverage that can be achieved by other technologies such as those provided by Bluetooth or Wi-Fi.

This approach was particularly appealing to operators who can offer both cellular access and wireline access based on DSL or cable to their customers, so that their subscribers could potentially access their fixed and mobile services with the same dual-mode device. It then comes as no surprise that British Telecom, at the time when it still owned its cellular operations arm, BT Cellnet, now branded as O_2, part of Telefonica (the Spanish operator, with fixed and mobile operations across Europe and Latin America), was the first company that started to actively investigate possible solutions and drive standardization of this technology.

Some time later, a consortium of companies, under the leadership of BT and other industry players, developed the Unlicensed Mobile Access (UMA) technology. This consortium was initiated in January 2004. The result of this work was the publication of an open set of technical specifications for extending mobile voice and data GSM/GPRS services over unlicensed spectrum technologies (including both Bluetooth and Wi-Fi). When this result was achieved, this alliance of companies encouraged these specifications to be adopted by 3GPP. In 2005, when the UMA technology was finally adopted by the 3GPP, it was renamed to Generic Access Network, or GAN. The terms UMA and GAN are now used interchangeably to refer to the same technology, and undoubtedly UMA will remain in the technical dictionary of many in the industry for quite some time, although the standard only speaks of GAN.

Overview The UMA/GAN essentially extends the GSM/GPRS services over the Bluetooth or Wi-Fi radio interfaces, and in general over any IP network handled according to GAN specifications, through a blend of VoIP technology and UMA/GAN-defined tunneling and signaling protocols. As we already mentioned, UMA/GAN is a GSM-specific technology and therefore can be used only in conjunction with the GSM/GPRS core network, in the profile defined by 3GPP to serve the GSM/EDGE Radio Access Network (GERAN), i.e., via the A and G_b interfaces. By supporting GSM voice and data services over Wi-Fi- or Bluetooth-based IP access, UMA/GAN provides a logical extension to the existing 3GPP systems, allowing operators to realize all the benefits of FMC. UMA/GAN technology aims also at allowing seamless handover between wireless local area networks running in an unlicensed spectrum, such as Wi-Fi or Bluetooth, and wide area cellular networks using dual-mode, GAN/UMA-capable mobile phones. Again, the wireless local area component of GAN/UMA may be based on any technology supporting IP connectivity, as specified in 3GPP TS 43.318 [80].

The traffic to and from a GAN/UMA terminal is routed over a GAN/UMA-defined interface to a GAN controller (GANC) (or a UMA network controller [UNC], in the legacy terminology), which appears to the GSM packet core as an ordinary base station controller (BSC). The GANC main function is to convert GAN signaling and media to a regular GSM voice call made over the GSM A interface. The GANC also converts the GAN data channel into a regular G_b interface–based GPRS packet data bearer service routed to the SGSN. Figure 5.6 clarifies the way GAN operates at a high level.

Under the GAN definition, the *local area* wireless network may be based on unlicensed spectrum technologies such as Bluetooth or Wi-Fi (IEEE 802.11)—or WiMAX

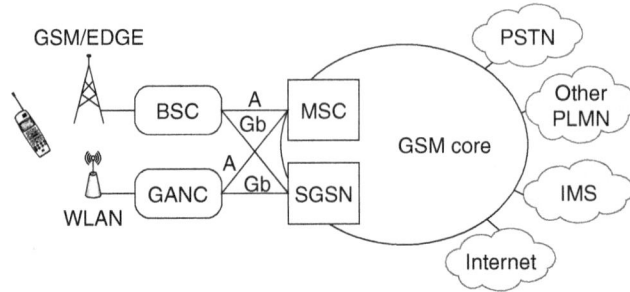

Figure 5.6   GAN high-level system view and concept

(IEEE 802.16) if the latter is used in *unlicensed* spectrum. The *wide area* cellular network may be based on GSM/GPRS or UMTS. Operators running a 3G network only, however, may not support in their core network the *A* and *Gb* interfaces that are necessary to connect the GANC to the core network. Thus GAN needs to evolve to support the UMTS-defined *Iu* interface so that 3G-only operators can use GAN without the need to deploy the 2G GSM-specific *A* interface in the MSC and the 2G *Gb* interface in the SGSN. By evolving to support the *Iu* interface, GAN could also take advantage of the enhanced performance and services offered by the 3G core network.

Technology and Architecture   The TSG GERAN in 3GPP defines and maintains the architecture (also known as stage 2, in this standards body's parlance) and protocol-level (also known as stage 3) specifications of "Generic access to the A/Gb interface," in 3GPP TS 43.318 [80] and 3GPP TS 44.318 [81], respectively. The GAN standard architecture is represented in Figure 5.7, where the various components of GAN are identified. The GANC and the GAN-capable mobile station (MS) are the new elements

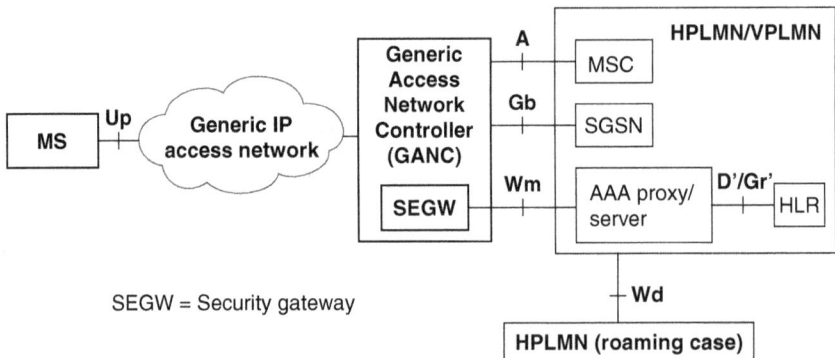

Figure 5.7   The GAN architecture

needed to introduce GAN features in an existing GSM network. Other elements depicted in Figure 5.7 belong to a standard 3GPP system architecture and the 3GPP/WLAN interworking architecture specified in 3GPP TS 23.234 [82].

GAN Interfaces   A GAN-capable MS needs to include, in addition to the existing GSM capabilities, the ability to connect via an IPsec tunnel to the GANC security gateway (GANC-SEGW) over the GAN-defined U_p interface. The setup of this connectivity involves the Internet Key Exchange (IKEv2) [83] protocol to set up the security association. The EAP-SIM [84] (for terminals supporting SIM only) or EAP-AKA [85] (UMTS AKA for terminals supporting USIM) procedure is started as a result of this exchange. The IPsec tunnel between the MS and the GANC-SEGW is then used to secure both signaling and user traffic between the MS and the GANC.

Another interface the GANC-SEGW needs to support is called W_m interface. The W_m interface is part of the 3GPP WLAN interworking architecture and is terminated at the 3GPP AAA server/proxy to perform user authentication at the IP access layer.

GAN Modes of Operation   A GAN MS supports four modes of operation:

- **GERAN-only**   Only cellular networks are used.
- **GERAN-preferred**   Cellular networks are used if available; otherwise, the GAN access is used.
- **GAN-preferred**   Use GAN if available; otherwise, use the cellular network.
- **GAN-only**   Use GAN only.

In all fairness, the nomenclature of these modes does not fully represent their functionality, as the GERAN mode can also be used over a UMTS network (not exactly based on GERAN). However, leaving this minor inconsistency aside, the modes themselves are fairly descriptive.

A GAN-compliant handset will scan for GSM (or UMTS) upon activation to determine its location area. This allows assigning the MS to the most optimal GANC, setting the correct charging information, and selecting the GANC in the visited network (if available) while roaming. When the MS is roaming, either the U_p interface may need to be extended all the way to the home network providing GAN service, or the visited network may provide the GANC, if a roaming agreement exists.

The discovery of the GANC address in a visited network is performed by first attaching to a "provisioning" GANC in the home network, and then obtaining the Fully Qualified Domain Name (FQDN—an identifier that can be resolved to an IP address using the DNS) or the IP address of the visited network GANC. The visited network identity is discovered when the GSM handset scans the network during activation. When a visited network GANC is used while roaming, the subsequent user authentication via the GANC-SEGW in the visited network entails using the 3GPP AAA proxy in the visited network to connect to a 3GPP AAA server in the home network, over the W_d interface.

Circuit and Packet Services User Plane   For circuit-switched services, the GANC performs the following user plane (U_p) functions, depicted in Figure 5.8, which represent the GAN CS user plane protocol stacks:

- Termination of the U_p IPsec tunnels, which carry the AMR/RTP/UDP/IP VoIP packets from the MS. In the protocol stack in Figure 5.8, these VoIP packets are using the "remote IP layer," which is the IP stack logically associated to the inner IP packets of the IPsec tunnel's virtual network interface.
- Termination of AMR/RTP/UDP/IP and framing into AMR before sending voice frames over the *A* interface using the necessary encoding:
  - If Transcoder-Free Operation (TrFO, specified in 3GPP TS 23.153 [185]) is not supported, then transcoding between the AMR and PCM needs to be performed for correct operation over the *A* interface.
  - If TrFO is supported, no transcoding is necessary at GANC, unless a common codec cannot be negotiated with the remote MS or transcoder.

For Packet Domain services, the GANC terminates the IPsec tunnels over the U_p interface and relays packets over the bearers of the G_b interface toward the SGSN.

Circuit and Packet Services Control Plane   On the control plane, the GANC (and SEGW) provides:

- Termination of the U_p IPsec tunnels
- MS/GANC mutual authentication, via IKEv2+EAP-SIM/AKA, and the support of the authentication procedures subset of the W_m interface specified in 3GPP TS 29.234 [186]

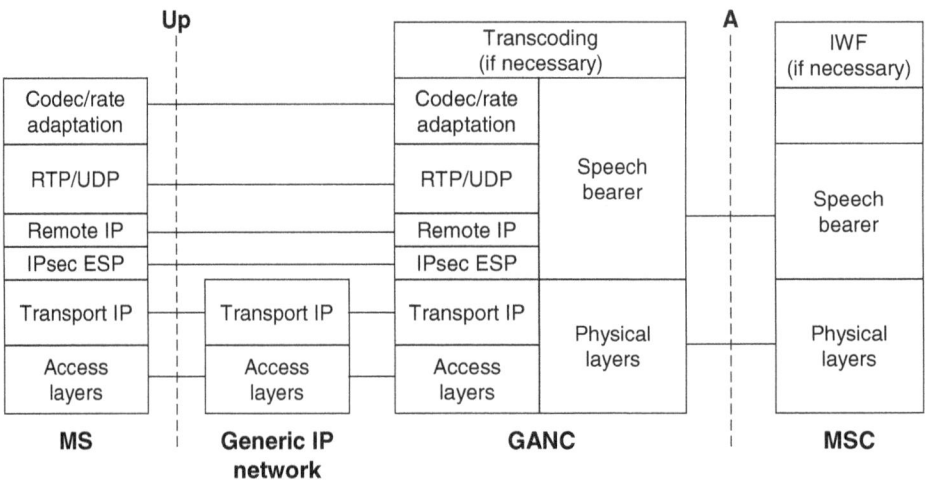

Figure 5.8   GAN CS user plane protocol stacks

- Transparent transfer of GSM/GPRS Layer 3 signaling messages between the MS and GSM/GPRS core network
- Registration for GAN access
- Providing GAN system information
- Establishment, administration, and release of control and user plane bearers between the MS and the GANC
- Support for paging, handover, and PS handover procedures

As an example of the control plane functions in GAN, the control plane protocols for the case of circuit-switched services are considered in Figure 5.9. This figure shows that the signaling transport service provided to the *A* interface by the BSSMAP [187] and the SS7 SCCP is replaced by the U_p encapsulation service provided by the IPsec tunnel between the GAN MS and the GANC. The reliable transmission service offered by the TCP protocol above the "remote IP layer" emulates the reliability offered by the SS7 SCCP layer. Finally, the Generic Access Resource Control (GA-RC) protocol plus the Generic Access Circuit Switched Resource layer (GA-CSR) offer signaling transport services similar to GSM-RR (Radio Resource) and GSM-RRC, used to transparently relay GSM Mobility Management–, Call Control–, and Supplementary Services–related signaling between the GSM MS and the MSC.

The descriptions of the user plane and control plane show that GAN encompasses quite a complex protocol set, essentially providing GSM PS and CS services emulation

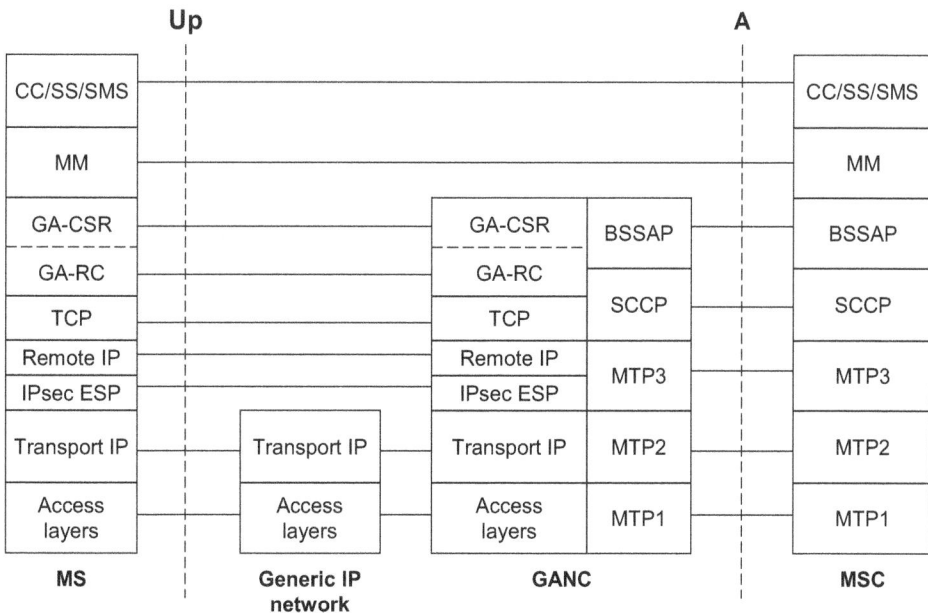

Figure 5.9   GAN CS control plane protocols

over a packet transport. Needless to say, the reliance on the GSM core has represented both the strength and the weakness of this solution. The following section evaluates the pros and cons of UMA/GAN, thus providing an insight on the GAN applicability to real-life deployments.

A Successful Standard or a Dead Evolution Branch?   While it is undeniable that many mobile operators (such as the ones quoted in Table 5.3—namely Orange, BT, T-mobile U.S., and Telecom Italia) have launched GAN services, it is still unclear whether GAN is bound to be a long-term success story or a short-term, gap-filling solution selected mainly for a time-to-market advantage and used only until IMS solutions become prevalent.

Indeed, while both UMA/GAN and IMS- and SIP-based FMC solutions are competing to enable similar user experiences and address the same needs, GAN is rapidly gaining a foothold in today's commercially available offerings. However, we should not rush into judgment and extrapolate these early successes to attempt prognosticating the future general adoption of GAN solutions.

As usual, answering the question of which approach will eventually be more successful is not easy, and in fact we shall not attempt it, focusing instead on providing an accurate analysis of the technology with its business model and highlighting its positives and negatives.

UMA/GAN Limitations   As we have already mentioned, UMA/GAN targets only GSM networks and does not provide a solution for 3GPP2-defined CDMA networks, comprising close to 30 percent of worldwide deployments. 3G-only networks such as the ones deployed by operators like "3" in the UK and Italy also cannot use GAN, since GAN by definition today does not support the I_u interface. Finally, fixed network operators wishing to roll out GAN-based FMC offerings have no option but to surrender subscriber control to GSM 2G operators, which tends to limit their flexibility in partnering and creating unique service offerings.

While 3GPP is in the process of enhancing GAN to offer 3G network operators the I_u interface they need, this enhancement will not address the problem of GAN subscriber ownership. Fixed operators entering partnerships with GSM carriers and wishing to retain a certain degree of subscriber control will have to establish the ability to terminate GAN traffic traversing through their GSM partner networks. For fixed operators, deploying MSC/GMSC and SGSN/GGSN functionality, and as such becoming GSM core network operators, may be possible in theory, but in practice it is impractical and difficult to justify from a business case and time-to-market perspective. Besides, such a setup would

TABLE 5.3   UMA/GAN Service Offers Launches in Europe

Operator	Date of Launch	Name of Service	Country
BT	Fall 2005	BT Fusion	UK
Orange	September 2006	Unik	UK, France, Spain, The Netherlands, Poland
Telecom Italia	September 2006	Unico	Italy
Telia Sonera	August 2006	Home Free	Finland, Sweden

require the establishment of new roaming arrangements with other 3GPP operators, which per se is a significant effort. In most cases this leaves wireline operators no choice but to fully outsource GAN-based FMC solutions to the cellular operator they partner with, and be content with back-end (for example, billing, Web services) ownership.

GAN limitations do not stop here. Fixed operators who have deployed GAN in partnership with a GSM operator that operates the GAN need in any case to route their fixed-line calls through a partner's GSM network core, a less-than-desirable scenario, especially in view of the deployment of an IMS core, which was designed to enable complete offload of the voice traffic from the cellular CS core. In general, the necessity to always route GAN traffic over the GSM core, even to or from the user in the home zone connecting through unlicensed non-GSM access, results in limiting cost benefits for operators (at least on the core-network-side utilization)—one of the main drivers behind FMC solutions deployment.

Furthermore, the lack of support for many PBX features (such as IP Centrex) has ruled out the adoption of GAN in enterprise environments, prompting the introduction of other types of FMC solutions not described in detail in this book, but mostly based on extending PBX capabilities in a proprietary manner.

Finally, the GAN-defined VoIP traffic generated by subscribers in the home zone is turned into a circuit-switched TDM voice by the GANC. The resulting possible "double"[11] encoding conversion, however wasteful, is necessary to offer the *A* interface to an MSC to support seamless handover, one of the main properties of a well-implemented FMC solution. Note that in principle it is possible to offer a true end-to-end cellular VoIP service over the GPRS domain of the GAN (assuming that QoS and bandwidth issues were somehow addressed), but then it would be impossible to hand over calls to the cellular network, as VoIP today is not commonly offered in 3GPP systems PS domain.

UMA/GAN Benefits   Our discussion in this section so far has focused on the potential limitations of GAN. In fact you might justly wonder how GAN could have been so successful with such a vast number of shortcomings. The answers become evident if one understands GAN focus and analyzes significant benefits this technology has to offer to 3GPP operators.

First and foremost, GAN looks like a logically straightforward extension of GSM service offerings, supporting all GSM supplementary services and providing comparable voice quality, which makes for a truly seamless user experience. GAN is also easy to deploy, especially for wireless operators, as the GSM core infrastructure would treat it as just another access method along with UTRAN and GERAN (that is, aside from the need to manage the GAN network and collect additional charging information that needs to be handled by the billing subsystem).

Finally, commercial implementation of a GAN-based FMC solution, which essentially requires the installation of only one new network element, GANC, is far less risky and quicker to the market than a full-blown IMS infrastructure (yet to be fully defined, integrated, and tested in most operators' networks).

---

[11] AMR to PCM and then back to AMR, if the remote end in the call is a GSM or GAN MS, or even to analog if the remote end is a traditional analog phone.

GAN-based Convergence   In this section we will consider practical aspects of the implementation of GAN-based convergence solutions.

Wi-Fi Access   With a GAN-based GSM/Wi-Fi dual-mode handset, it is possible to place and receive calls over the cellular network or over the broadband access network supporting the GAN access. As there are no specific requirements on the Wi-Fi access point or type of broadband service used by the GAN subscriber, the Wi-Fi access options may include a residential or small office DSL, cable, or fiber broadband connection with the appropriate CPE, a Wi-Fi hotspot, or a college campus WLAN deployment. Upon entering a Wi-Fi hotspot, or a home zone Wi-Fi coverage area at home or at a small office, the handset gains wireless access over the Wi-Fi network. This may require configuring the GAN terminal with the appropriate SSID of the Wi-Fi networks and possibly other parameters as described in the preceding chapter.

For example, in some cases, the Wi-Fi access points may implement 802.1x security and require authentication for wireless access, and in this case support for an EAP exchange using EAP-SIM or EAP-AKA may be needed to provide access control. The operator may also use a AAA server to implement a set of policies governing the access over nonlicensed networks based on location, time of day, presence, or other parameters.

Once the dual-mode GAN terminal is connected to the Wi-Fi network, an IPsec tunnel is established between the GAN UE and the GANC. The GANC includes a security gateway authenticating the terminal and providing access control. This authentication also allows the AAA server to check the location of the user based on the correlation with information derived from the network access authentication and the IP address. This information is useful for charging purposes and also for allowing location-based and user-profile-based access control. Note that this level of authentication is different from authentication used to gain Wi-Fi access.

The peculiarities of Wi-Fi-based GAN are generally limited to the ones described already, as the GAN architecture in principle varies little between different unlicensed wireless access technologies used to obtain IP connectivity. Similar considerations also apply to Bluetooth access, the second most popular GAN mode of operation in the home environment after Wi-Fi, which is described in the next section. One significant difference in Wi-Fi and Bluetooth in practical GAN solution implementation is that Wi-Fi proves to be quite uniquely positioned for public hotspot deployments, whereas Bluetooth is not really applicable in that environment.

Bluetooth Access   Despite proliferation of Wi-Fi, the initial deployments of GAN have taken place with the GSM/Bluetooth dual-mode terminals, as the availability of GSM/Wi-Fi terminals until recently has been quite limited especially in mid-tier consumer price segments, making positioning GAN solutions with mainstream subscribers difficult. Another issue with dual-mode GSM/Wi-Fi handsets has been (and still is) their poor battery life while in Wi-Fi access mode, sometimes measuring at 50 percent that of GSM, especially in standby, where constant AP handshaking is necessary.

In these early deployments, Bluetooth GAN terminals based on the regular Bluetooth-equipped cellular phones, needing little or no modification to support UMA/GAN clients, were logically selected to fill the gap. An example of such early launches was the *BT Fusion* service, initially named "Bluephone," which started with Bluetooth GAN handsets and only recently expanded to Wi-Fi.

While offering these and a few other advantages, the deployment of Bluetooth solutions still requires a *Bluetooth hub* in the home zone, grooming the GAN traffic toward the broadband access network. Bluetooth hubs have a nasty habit of "seizing" the handset, often without regard for subscriber preferences, as the support for specific Bluetooth profiles in handsets varies significantly between manufacturers. Thus GAN Bluetooth hubs were frequently found "dueling" with Bluetooth headsets, or in the opposite extreme case the service required the subscribers to manually switch over the handset to and from headsets and hubs. The situation got so bad that in some cases the only available solution for operators was to provide the GAN users with a corded headset and restrict them from using Bluetooth headsets and other devices while in the home zone.

Let's now continue with an overview of another vertically integrated convergence solution, based on the *femtocell* concept.

Femtocells

A *femtocell* is a modern, smaller scale, reincarnation of a nano- and picocell technology first introduced in the previous decade. Specifically, nanocells and picocells were initially brought to market in the late 1990s by companies like Nokia, Motorola, and Ericsson. Unfortunately these products were not well received. Among the reasons at the time were limited backhaul bandwidth, high equipment cost, and poorly chosen deployment strategy—a typical combination of factors for a technology ahead of its time.[12]

Femtocell solutions nowadays are used to convert traffic to and from standard cellular handsets in close proximity (typically up to 100 meters) and carry it over IP, reducing to a bare minimum the air portion of the cellular traffic and relieving operators from the necessity to deploy expensive wide area cellular sites.

Essentially a femtocell is a miniaturized base station with the following characteristics:

- Radiated power in the single-digit milliwatt range
- Capacity between four and ten simultaneous active calls
- Backhaul based on IP over broadband links
- Range around a hundred meters

Femtocells are typically designed to be installed in residential home zones or small offices. They are designed to provide the same service as any other base station to subscribers using standard cellular handsets.

---

[12] Limited acceptance of IP at the time as a common Network layer communication technology for voice and data, and also for wireless access backhaul, may have also had its influence. The introduction of mass-market IP broadband access contributes to making deployment of femtocells quite practical and economical these days.

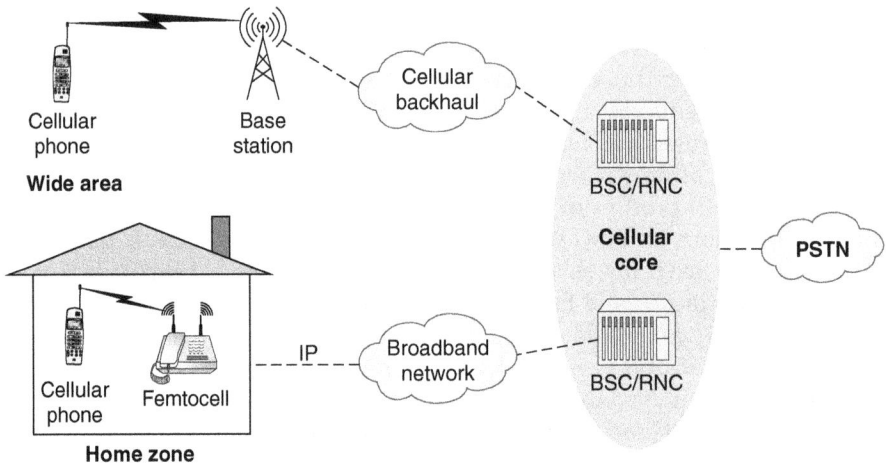

Figure 5.10   High-level femtocell solution architecture

The value proposition behind femtocells is similar to that of the dual-mode solutions, addressing *cost, capacity,* and *coverage* issues for the operator—allowing the reduction of both CAPEX and OPEX—and in some ways *convenience* issues for the subscribers. Thus, at least in theory, this technology lends itself well to enabling many elements of the converged communication experience. We must note, however, that femtocell technology is difficult to apply to enable truly converged FMC solutions, as it does not support[13] multimode (cellular plus some alternative access technology) terminals and access-independent communication, principally addressing the issues of limited indoor area coverage, and of the offload of traffic from congested licensed spectrum, using the same technology as wide area cellular networks.

Architecture   A typical femtocell solution consists of two main component classes: the actual CPE femtocell and a backhaul component, usually a gateway or concentrator, converting femtocell output to media and signaling understood by the rest of the infrastructure. The functionality of such elements is somewhat similar to that of GANC.

Figure 5.10 provides a high-level view of a typical femtocell solution.

Today's femtocell backhaul architecture options include

- A Modified RNC approach constituting tunneling an I_{ub} interface to a modified cellular radio network controller (M-RNC)
- A concentrator or controller approach supporting connectivity to a gateway rather than a direct RNC connection
- A collapsed stack femtocell

---

[13] At least at the time of this book's writing.

- A UMA/GAN femtocell
- An IMS femtocell

Next we analyze and compare these approaches.

Modified RNC   The Modified RNC[14] option simply provides a way for a femtocell to connect *directly* to the operator's UMTS RNC (or BSC in CDMA and GSM networks). The standard I_{ub} transport protocol is emulated[15] in this architecture to encapsulate voice traffic media and signaling in IP, tunnel it over a broadband connection, and terminate it at an I_{ub}-compliant Modified RNC.

The femtocell traffic is different from that of a regular base station, in that the RNC may terminate E1/T1s from the Node-Bs and may not support the IP transport option; therefore, the RNC must be *modified* to accept IP–based I_{ub} (defined in [86] [87] [88] [89] [90] [91] [92] [93]), hence the name: Modified RNC (M-RNC). Even with this necessary modification, this solution is the one introducing the fewest changes in the existing cellular infrastructure and, in many cases, is the easiest to deploy.

Note that one of the potential drawbacks of this solution is the RNC infrastructure scalability, as the existing RNCs are designed to support a limited number of base stations and may not be able to accommodate the conceivably millions of femtocell connections that could result from the "one femtocell in every home" campaign. So, different options are required to support these femtocells, also to take into account the fact that IPsec termination may be required to secure the path between the femtocell and the RNC. This is provided by the RNC concentrator.

I_{ub} Concentrator   The I_{ub} concentrator approach was invented to address the drawbacks of Modified RNC solutions, specifically low scalability and the need to modify many RNCs in the existing infrastructure. As the name implies, the femtocells in this option connect to a highly scalable I_{ub} concentrator terminating individual I_{ub} tunnels (potentially protected by IPsec), converting it to the format friendly to the standard RNCs, as in Figure 5.11.

Some femtocell concentrator vendors even go as far as building controller functionality into their products, allowing to connect them directly to the core. That approach enables them to offload RNCs when subscribers are using femtocells en mass, potentially at data rates and for durations and applications not common in the wide area, which would make the traditional traffic model used to size RNCs no longer valid and so congest existing RNCs' capacity.

---

[14] In this section we will refer to the interface used in UMTS to connect a UMTS Node-B to the RNC, but the equivalent interface used to connect base stations to their controllers in other technologies can be assumed *mutatis mutandis* (that is, applying the appropriate terminology applicable for the different technologies). In 3GPP technologies, we have chosen to focus on UMTS, as operators will focus on this technology, while not discounting 2G technologies, in the future evolution of their network.

[15] The I_{ub} interface is used in today's cellular networks to connect UMTS base stations (Node-Bs) to UMTS RNCs.

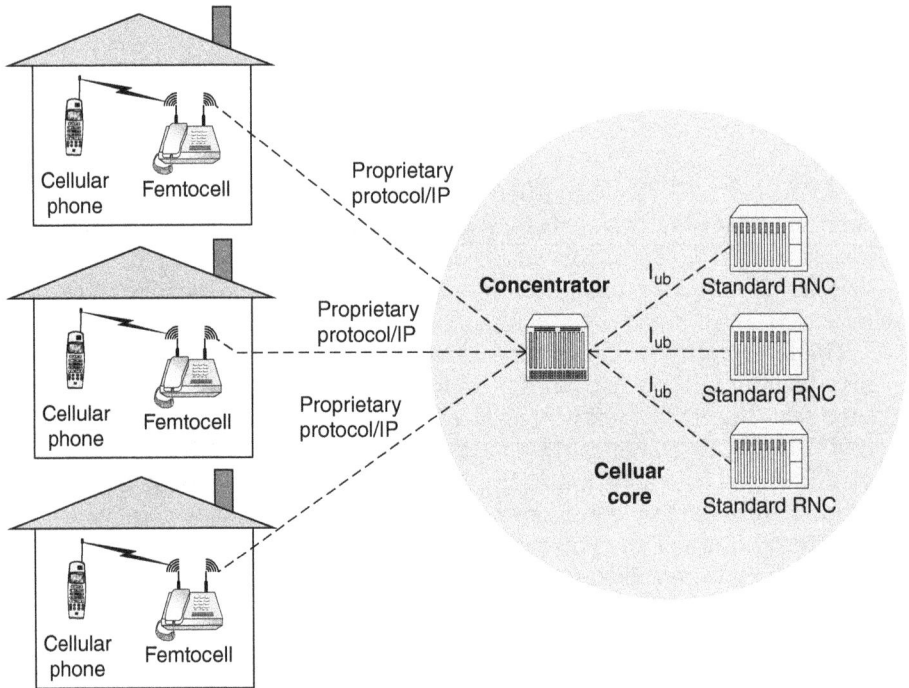

Figure 5.11 I_{ub} concentrator

On the downside of the I_{ub} concentrator approach, the operators will need to introduce new elements into their networks, in most cases supporting proprietary protocols, effectively creating a "network within the network," as Figure 5.11 clearly demonstrates.

Collapsed-Stack Femtocell   The solutions based on *collapsed-stack* femtocells, also called femtocell *access points,* shift the intelligence from RNC and even SGSNs or PDSNs (in the case of PS traffic) into femtocells. Femtocell access points are typically designed for use by larger enterprises and scalable to support hundreds of users. The femtocell access points often incorporate radio network controller (RNC) functionality instead of relying on the server in the core network. Such "intelligent" femtocells are then connected directly to the core network, as shown in Figure 5.12.

This approach, essentially moving a large chunk of cellular core in the subscriber's premises, has some interesting benefits, most notably allowing for *localized switching*. Smart femtocells in this solution can even be used to replace some PBX functionality in an enterprise environment or even potentially augment advanced home telephone systems.

The unfortunate reality of a collapsed-stack femtocell is that operators might be reluctant to entrust so much of their infrastructure intelligence into the hands of a common subscriber, especially if not only the RNC, but also the SGSN or other functions, are

RNC/GSN - femto

DCH DSCH FACH	RLC RRC	SCTP/GTP-U
Framing (MAC)		UDP/IP/IPsec
UMTS PHY		L2
RF		L1

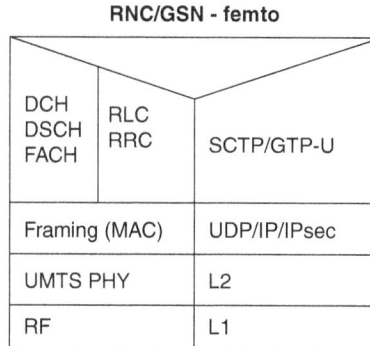

Figure 5.12 Intelligent femtocell protocol stack

collapsed in the node. On the other hand, collapsed-stack femtocells are capable of offloading many functions traditionally associated with the core infrastructure to customer premises, effectively distributing the load on the network and its processing power, and enhancing system scalability.

Needless to say, in the case functions such as the SGSN are supported by femtocells, the MAP or IS-41 interfaces need to be exposed, and this involves the need to concentrate this signaling toward the HLR in the core network (HLRs are typically not scalable enough to support millions of simultaneous SS7 connections).

GAN Femtocell   What is GAN doing in the femtocell section?—you may ask. Didn't we establish the femtocell technology as an alternative to dual-mode solutions? To understand the logic here, you may need to revisit the UMA/GAN section to discover that despite UMA being originally developed to support GSM traffic over *unlicensed* spectrum, its concept and technology can be successfully applied to *licensed* air interfaces as long as IP is used for transport and broadband backhaul is available.

The main distinction between GAN dual-mode and femtocell implementations is that with the latter, a GAN client is supported in femtocell rather than in the handset. As shown in Figure 5.13, in this approach the GAN U_p interface encapsulating A and G_b, is established between a GAN-compliant femtocell and GANC in the operator core infrastructure, which converts GAN signaling to regular GSM traffic.

GAN femtocells make a lot of sense. For one, this approach is in large part standards based, with all the typical benefits of standards-based solutions. Operators embracing GAN now have both dual-mode VoIP and femtocell deployment options and therefore can more finely tailor their solutions to specific markets and subscriber segments. On the downside, this approach is only available to GSM operators and (as typical for GAN), unlike the IMS approach described next, does not offload the user traffic from the cellular core.

IMS Femtocell   The final femtocell deployment option goes one step further than the other approaches by converting cellular traffic to VoIP and "never looking back."

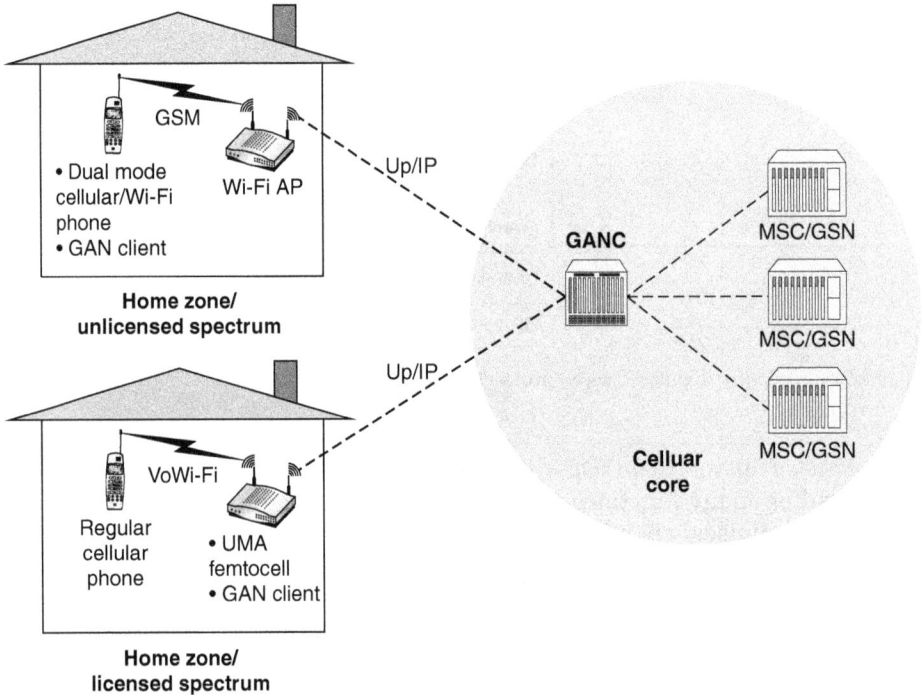

Figure 5.13   GAN femtocell vs. Wi-Fi

That is, the traffic from the femtocell is not converted back to cellular but rather processed as VoIP signaling and media in the operator's IMS core (or, for that matter, in any correctly implemented VoIP core). With a stretch, this approach can be regarded as a particular case of the collapsed-stack femtocell. Indeed, it supports either native IMS-based communication on the PS domain access or interworked IMS communications via an MGW functionality in femtocell, controlled by an MGCF in the IMS core via the M_n interface.

Figure 5.14 will help you to visualize the benefits of this solution, providing familiar capability to offload traffic from both wireless WAN and cellular core. On the downside, this solution is only suitable for operators with a deployed IMS infrastructure, so it may not be viable for the majority of carriers out there, at least in the short term.

Femtocell Pros and Cons

Femtocells are relative newcomers to the market and have not yet proven themselves with either operators or mainstream consumers. As with any new technology, femtocells have their pros and cons that must be considered in detail by operators and vendors to ensure successful market introduction and acceptance. In the following section we are going to analyze these technology challenges and benefits.

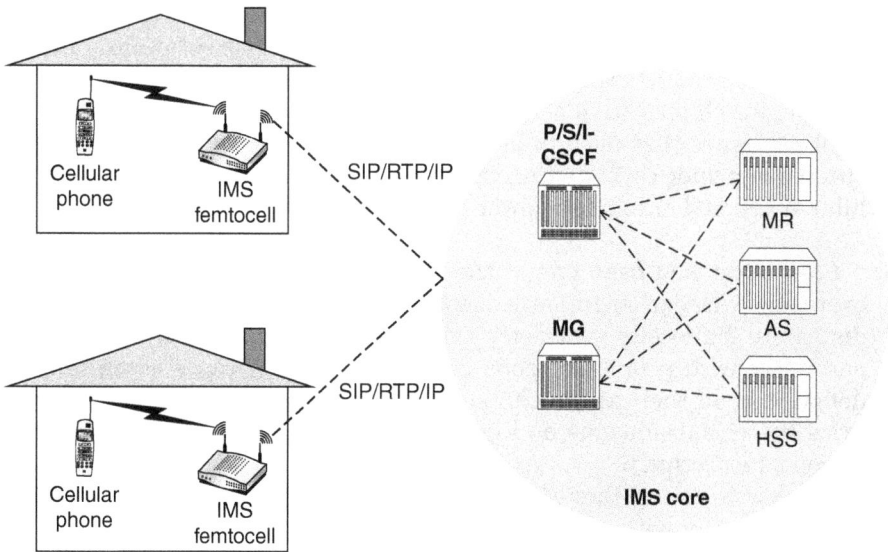

Figure 5.14   IMS femtocell

Femtocell Limitations   Let's first consider femtocell challenges.

Standardization   Despite the long history behind the whole mini–base station paradigm, as of today there are still no real standards associated with femtocells. The femtocell commercial offerings are therefore evolving as proprietary solutions, sometimes partially compliant to the existing 3GPP and 3GPP2 standards governing traditional cellular architectures.

As with any proprietary solutions, operators and consumers alike stand to benefit from their quicker time to market and more diverse sets of functionality. On the other hand, the femtocell manufacturers and suppliers will not be able to realize the economies of scale associated with standardized solutions, a situation that is in turn bound to affect the femtocell costs already identified as one of the potential roadblocks to widespread deployment.

Spectrum   Unlike Wi-Fi access points, femtocells by definition operate in *licensed spectrums* using standard cellular air interfaces and protocol technologies such as GSM, UMTS, or CDMA 1xEV-DO. As such, femtocells may only be deployed in areas where operators obtain spectrum licenses. That may make femtocell deployment problematic, especially in the geographical border areas.

Also the subscribers will need to be educated on the difference between solutions operating in licensed versus unlicensed spectrums (e.g., femtocell vs. Wi-Fi hotspot), and be prohibited from operating their femtocell-handset combinations in areas where no licensed spectrum is available or where their operation is undesirable to carriers, such as in a different country or continent. To enforce such restrictions and prevent subscribers from turning femtocells into portable base stations, manufacturers often equip their products with GPS receivers and appropriate software capable of managing[16] femtocell operation according to location.

Other issues that need to be addressed for a typical device operating in a licensed spectrum include coexistence, interference, and communication with the rest of the cellular WAN, and automatic power control on par with regular cells.

Cost   Cost had been one of the most serious hurdles the industry was unable to overcome with earlier miniaturized base-station forays. Operators naturally expect to distribute femtocells—correctly viewed as cellular infrastructure extensions—to the end users with poor home-zone coverage, in many cases essentially addressing the deficiencies in their service. In such environments many operators identified the top price before subsidies as no higher than around $300 or €200, in the Americas and Europe, respectively.

It's worth noting that the impact of femtocell cost on carriers' pricing decisions changes widely with a given operator's distribution model. When femtocells are distributed by a carrier as a remedy for coverage gaps, the costs are naturally expected to be absorbed by the carrier itself. In some cases, however, carriers may require (or allow) residential or enterprise subscribers to acquire or lease the equipment—similar to the current cable TV model. Finally, carriers perhaps may rent femtocells to organizations or individuals or share them to provide service in public locations—as with current tower-sharing arrangements.

Integration Issues   Femtocells may certainly help fixed-to-mobile substitution, making it easier (and cheaper) for the subscribers to start using only their mobile phone (and service) for all of their communication needs. However, when it comes to FMC, femtocells may be difficult to integrate with the existing fixed services in use by subscribers, such as office PBX-based systems in enterprise environments or residential multi-handset analog cordless systems or VoIP phones. Therefore, femtocell-based solutions are often likely to stay a step behind other FMC approaches such as dual-mode cellular/Wi-Fi methods, in terms of convenience and usability for many market segments (for example, for subscribers interested in using multiple types of devices in FMC family plans).[17]

To overcome these limitations, femtocell solutions will have to incorporate Wi-Fi access points and other functionalities that may make them more complex and drive their costs even higher, which in turn affects subscriber acceptance.

Benefits   After reading the preceding section, you might think that femtocells are doomed even before being born. The reality, though, proves this conclusion premature,

---

[16] Read: restricting.

[17] See examples in the next chapter.

especially when applied to specific subscriber segments and markets. Operators looking for ways to deploy femtocells must therefore carefully select the deployment strategy and consider femtocells as an integral part of their existing infrastructure.

Femtocell technology benefits for equipment manufacturers, operators, and subscribers are significant and diverse. By comparison with other FMC approaches, femtocells allow subscribers to use their *existing* devices to achieve many of the advantages promised by FMC, such as excellent indoor coverage and home-zone pricing without the typical limitations of dual-mode solutions. Such limitations might, for example, include the need for the users to upgrade to new, potentially more expensive dual-mode terminals, coupled with limited terminal selection.

Femtocells' positive impact further increases as 3G solutions proliferate. This is mainly due to 3G systems' higher frequencies, subsequently higher indoor signal obstacle path loss, and ensuing signal degradation with the increased number of indoor users of 3G service—an all-too-common scenario in urban areas.

Other benefits of femtocells include improved handset battery life compared to Wi-Fi or even the regular wide area cellular coverage in the home zone (due to lower required transmission power), improved ability to conditionally throttle the network load (both backhaul and radio interface), and last but not least, cheap on-demand capacity during certain times and events (such as trade shows, or mass entertainment and political events) where femtocells can be used to set up temporary "cellular hotspots."

Expanding the Femtocell Definition   Femtocells enable only partial FMC by using wireline broadband access as a backhaul solution, but do not alter the fixed and mobile services experience or user terminals. On the other hand, a femtocell may incorporate other residential and small business telephony and wireless functions such as

- Wi-Fi AP (to enable users in the home zone to continue using 802.11 devices[18])
- Terminal adapter (to act as a base for cordless phones)
- Broadband modem (enabling wireline connectivity)
- Cable TV set-top box (or rather the set-top box may support femtocell functions, depending on the point of view)

Such an integrated class of devices would allow operators both to offer indoor cellular connectivity and to support its subscriber landline and broadband services, helping to enable the coveted quadruple-play service offerings.

These devices can be used to enable simultaneous ringing on multiple handsets and numbers, conditional forwarding, and other FMC functionality traditionally associated with dual-mode solutions.

---

[18] Alternatively, femtocells may connect to an Ethernet port of an existing DSL/cable mode wireless router in the customer premises. In this case femtocell's support of 802.11 would be unnecessary.

Summary

In this chapter we have walked through the analysis of the various convergence methods available in the industry today to enable practical FMC service offerings. These methods have so far been presented mainly in the domain of the converged voice and data services applications. Along the way we analyzed main enablers such as IMS VoIP, GAN/UMA, and femtocells.

The following chapter takes the discussion one step further by analyzing practical aspects of FMC as a whole and paying special attention to the subscriber experience it enables. We will also explore new applications that are only possible in FMC systems as well as find new uses for the existing applications extended over converged networks.

6

FMC Impact on Services and User Experience

You cannot acquire experience by making experiments. You cannot create experience. You must undergo it.

—Albert Camus

. . . every experience is a form of exploration.

—Ansel Adams

A product or service enabling a unique and compelling user experience, creating value for the users, and allowing for stronger differentiation and value creation is a cornerstone of any successful operator and vendor strategy. In this chapter we focus on such experiences and use-case scenarios made possible by FMC technology and its business models. While exploring the use cases in detail, we also evaluate applicability of the technologies outlined in the previous chapters. At the end of the chapter we look at a few technologies related to how FMC helps to enable these use cases.

This exercise should appropriately put to the test everything you have learned thus far about convergence. It will also help you advance your practical understanding of FMC by exploring some real-world scenarios and seeing seamless mobility in action.

FMC Scenarios and User Experience

This section provides an overview of services and experiences enabled by various FMC technologies and solutions. We will start the discussion with the unique pricing and distribution models as well as converged billing and provisioning methods. FMC solutions based on these approaches are often referred to as *lightweight* FMC or *loose* convergence.

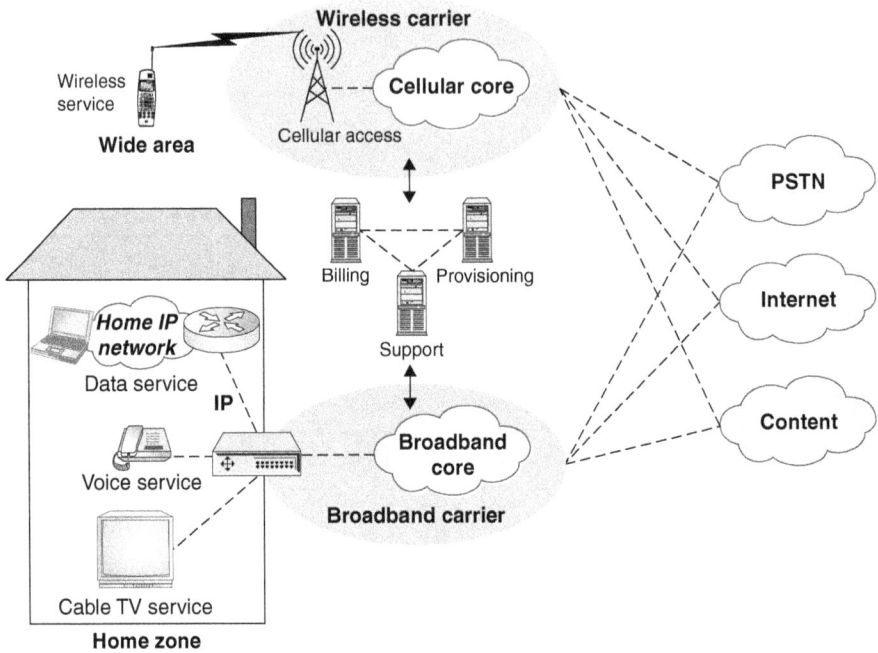

Figure 6.1 Loose convergence solution example

As we progress through the chapter, we introduce more complex and more tightly integrated solutions defined by the number of handsets and access identifiers, such as telephone numbers, supported per FMC services account.

Lightweight FMC

Loose or lightweight convergence can often serve as a safe first step in the operator's FMC rollout strategy. The techniques making this approach possible often enable operators to provide basic converged services without the risk and expense associated with the "full-blown" FMC solutions requiring new handsets and extensive core network modifications. Figure 6.1 pictures an example of such a lightweight FMC solution.

Back-End Convergence   Just like fixed-mobile *substitution,* which can be stimulated by imaginative pricing strategies such as home-zone pricing (e.g., subscribers pay nothing or a minimal flat fee—usually comparable with fixed calling—when calling from home[1]), some simpler forms of lightweight FMC can also be provided via clever use of a combination of pricing, billing, and distribution. We will address this method as *back-end convergence.*

---

[1] See the section "LBS" at the end of this chapter to find out how service providers can determine subscriber locations.

An example of a back-end convergence approach would be a certain flavor of a "quadruple play," a concept you should find familiar by now. The users of individual services offered by different providers will be presented with a plethora of incentives to migrate to a convenient bundle of four services such as broadband Internet access, media content, and both landline and mobile voice. Since this will entail an update to the subscriptions and changes to the homing of these individual services in the network, this approach does not technically require installation of a new convergence-specific CPE such as a femtocell, or the purchase of new handheld devices, or introducing the users to a new type of service. Of course new CPE may be required for transition, such as a new set-top box or a new DSL wireless router, but these are the normal nonconverged service-oriented devices.

3GPP View   Back-end convergence allows service providers to address many user needs without actually going into the expense of building fully converged solutions, but resorting instead to convergence of their existing back-end systems (hence the name: *back-end convergence*). The majority of back-end convergence solutions from the mobile operator perspective correspond with the 3GPP *Common Billing and Customer Care and Access Control and Charging* scenarios defined in the 3GPP TS 23.234 [82].

According to the first 3GPP scenario, the convergence between the Wi-Fi- and the 3GPP-defined cellular systems will only be represented by a single customer relationship. The customer receives one bill from the mobile operator for the use of both 3GPP and Wi-Fi services. Integrated Customer Care allows for a simplified service offering from both the operator's and the subscriber's perspectives. The security and authentication of the two systems may be handled independently.

The second scenario defines the solution where authentication, authorization, and accounting are provided by the 3GPP system. The security and authentication in this case are handled by the 3GPP system for all types of access to increase the "seamlessness" of the service and provide for a more consistent experience while using different access methods or in some cases end user devices.

Benefits   The main goals of the back-end FMC are to improve subscribers' experiences and provide subscribers with lower pricing, with 10–20 percent discounts to be expected. Various pricing models can also be applied instead of just flat discounts, depending on specific usage patterns the service provider seeks to promote. For example, the service provider may offer various discounts for both fixed and mobile calling in the home zone based on network load, time of day, and other conditions.

Converged distribution and support promise to enhance both pre- and post-sale user experiences for both organizations and individuals. The user satisfaction is improved from the get-go with one-stop shopping when the new user can sign up for all four services and purchase all the required components in one shot either online or in person in a customer service center.

The benefits do not stop here. The customer will receive support for all communications services from a single support department, with only one number to dial.

Common billing further improves user convenience. Indeed, a single bill providing itemized information on all communications services consumed by a subscriber does not require much marketing. Converged billing, however, still requires a significant effort on the core side to consolidate back-end systems and accounting processes, which might present difficulties for operators entering partnerships to provide different services.

Finally, back-end convergence makes for easy deployment of common network-based voice mail (VM) and contact list solutions. The unified voice mail can be more easily accessed from different, even nonconverged devices when both fixed and mobile services are provided by the same carrier or partnering carriers sharing (or even outsourcing to a third party) common VM.

Limitations We must note that while back-end convergence significantly improves subscriber *convenience*, with all its benefits it does not really address the capacity, cost, and coverage—the *C*s that drive FMC—being really only a "paper" convergence. Operators offering back-end convergence solutions will not be able to provide their subscribers a truly converged service experience. That is, the users of a back-end FMC would still have to rely on multiple devices with multiple access numbers to communicate over different access networks. Moreover, back-end convergence does not provide any mechanisms enabling handover between fixed and mobile communications.

The downside of loose convergence with regard to pricing is that the service provider *costs* are usually not affected (with the exception of potential modest savings on unified support and billing), which makes for a weak business case. This is one of the reasons that operators to date have been cautiously conservative, going after strategic objectives rather than the bottom line, when rolling out such offers on a broad scale. Examples of these strategic objectives could be retention of customers or customer acquisition in mobile or fixed environments, reaction to competitive pressures, or attacking a competitor in one area of strength.

Table 6.1 provides a view of the FMC requirements satisfied by back-end convergence and illustrates its benefits and limitations.

TABLE 6.1 Back-End FMC Features

Feature	Supported?	How?
Convenience	Yes	Single bill, single point of contact
Coverage	No	Services as well as core and access stay physically separate
Capacity	No	Services as well as core and access stay physically separate
Cost	Partially	Service provider cost can be affected only slightly by eliminating duplication of service billing and support personnel
Home-zone pricing	Yes	Through strategy but with weak business case
Single bill	Yes	Billing systems convergence
Common distribution	Yes	Common distribution
Converged support	Yes	Common tier-1 support team

TABLE 6.1   *(Continued)*

Feature	Supported?	How?
Fixed-mobile roaming	No	Services as well as core and access stay physically separate
Fixed-mobile handover	No	Services as well as core and access stay physically separate
Single number	No	Services as well as core and access stay physically separate
Converged multinumber	No	Services as well as core and access stay physically separate
Multimode device	No	Services as well as core and access stay physically separate
Converged VM	Yes	Through a common converged VM system
Converged contact list	Yes	Requires advanced VoIP phone
Find me/follow me	Yes	Possible in VoIP service and certain cellular services
Conditional forwarding	No	Needs special equipment

It must be noted that many of the back-end convergence limitations can be addressed by combining back-end convergence with one or more of the *conditional forwarding* techniques described in the next section.

Use Case: Back-End Convergence

After getting his MBA from the University of Stockholm, Hans Johansen has started a home-based business requiring much communication, both fixed and mobile. Therefore he spends a significant amount of time managing two separate accounts and deals with billing and service issues separately. Lately the charges from both of his service providers, Blue Sky Wireless and Green Earth Telecom, have been rising precipitously, so Hans decides to check out a new service promoted by a new company, ConvergeCom.

Hans is particularly intrigued by an advertisement promoting converged fixed (VoIP) and mobile (cellular) individual plans. He calls the operator's customer care and the switch is quickly arranged. The new plan, while providing the same landline service and an equal amount of cellular minutes at a flat rate (as Hans' current mobile plan), is 20 percent cheaper than what Hans used to pay separately for his landline and mobile services.

The next week, Hans receives the welcome package from ConvergeCom containing the following:

- VoIP router with built-in analog phone adapter (TA) with all necessary cables
- VoIP service manual and quick-start guide
- Advanced fixed VoIP phone with LCD display
- Mobile phone with chargers and cables

(continues)

- Service manual and quick-start guide covering both fixed and mobile service
- Service introduction letter highlighting common fixed and mobile features, including VM, contact list, and billing and support procedures

As Hans starts using the service, he is delighted with the convenience of a single voice-mail system and a contact list that is shared and synchronized between both his mobile and advanced VoIP phones. The contact list can also be backed up to a server in the network and restored remotely if Hans decides to switch to different devices.

At the end of the month a single bill comes. The bill provides itemized information on both services, and Hans pays it online in seconds. He is delighted. He only wishes the two services in the plan would be more converged. . . .

Conditional Forwarding   On-demand forwarding—or the ability to make incoming calls made to a certain number ring on another, user-defined number—can be *permanently* invoked on most modern voice communication systems via a Web interface or, for example, by dialing *72 on the *from* line and then entering the *to* number.

While on-demand forwarding has generally been popular with the subscribers, its use is limited mainly due to the need for constant user interaction (to manually turn it on and off) and insensitivity to location and other conditions. For instance, in most cases, to invoke or disable the service the subscriber must be physically present at the *from* location or have access to the *from* device (when forwarding from the mobile line). Also, traditional forwarding is static in nature and cannot be automated. (For example, the user cannot give this command: "Unforward when I am at home.")

Simultaneous ringing of multiple lines on incoming calls (often referred to as a "follow me" or "find me" service) is another service related to on-demand forwarding. It is commonly offered as a feature of most modern VoIP solutions and can be invoked from a Web interface. It allows the subscriber to select multiple numbers (fixed or mobile) that will ring simultaneously when the call is placed to a fixed VoIP number. When both mobile and fixed services are offered by the same operator or specific agreements are in place between collaborating fixed and mobile operators, it is even possible to provision the service to allow ringing on all selected numbers whenever the call is placed to any of the numbers on the account, both fixed *and mobile.*

Just like on-demand forwarding, simultaneous ringing has proved to be a valuable feature for both subscribers and operators (fewer lost calls = less lost revenue), but it also has significant drawbacks. For example, placing calls from one of the *follow me* numbers to a primary number results in a race condition and often drops calls, on top of potentially causing confusion to subscribers. To avoid such undesirable side effects, a subscriber has no option but to proactively manage the feature—for example, to remember to turn it on or off in particular situations—similarly to that of the on-demand or *conditional forwarding* (described next), which significantly decreases its usability—or dial a certain star code before calling the primary number. In another instance, when one of the numbers in the group belongs to a mobile phone which is off or out of coverage, the calls placed to a fixed number will be automatically deposited straight to a mobile VM without giving the subscriber a chance to pick up.

So after a few years the industry is coming to the realization that with its apparent limitations, on-demand forwarding is not going to cut it as a mass-market convergence service without major enhancements, some of which may be provided by *conditional forwarding*.

Conditional or automatic forwarding takes the user experience one step further by enabling the forwarding of calls between different numbers based on specific *conditions*, such as location, presence (see further in the chapter), or time of day—all without any user involvement. For example, the users may choose to forward all of the calls inbound to their mobiles to a fixed number only when they are present in the home zone. Other options may allow them to set up the service to automatically invoke forwarding of the office number to a mobile number after a certain time of day or when the subscriber leaves the office.

Conditional forwarding may be implemented via various mechanisms such as Bluetooth Hands-Free Profile (HFP) or Cordless Telephony Profile (CTP) notification or through proprietary solutions supporting some mechanism to automatically notify the network about changes in subscriber condition. Figure 6.2 provides an example of such a solution relying on Bluetooth HFP-based location notification. In this figure a conditional forwarding gateway is used to detect when the subscriber with a mobile phone enters the home zone and subsequently notify the infrastructure so that it can forward all mobile-bound calls to the subscriber's fixed number.

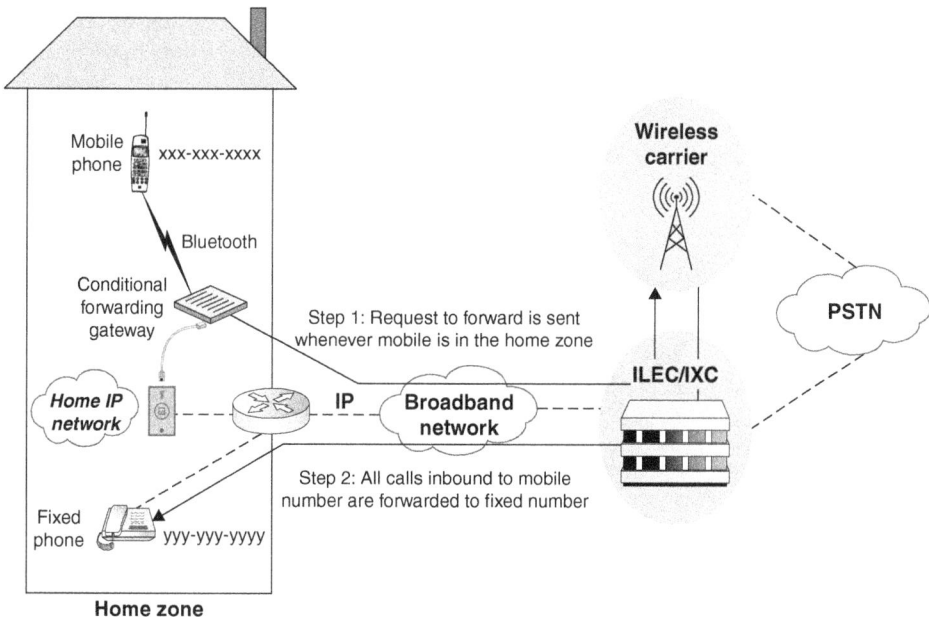

Figure 6.2 Conditional forwarding example

Note that the subscriber's location detection is based on the interaction with his or her mobile phone, and therefore will only work if the phone is powered on when the subscriber enters the home zone.

Conditional forwarding can also be applied to address the shortcomings of simultaneous ringing. For example, the *follow me* feature can be disabled automatically when the subscriber enters the home zone and his or her mobile phone is on.

If applied in this fashion, conditional forwarding essentially simulates an elementary ability to *roam* between fixed and mobile services and access networks (albeit only for incoming calls). The subscriber still has to switch between different devices to make calls over different access networks; however, *incoming* calls are routed through the appropriate network to the subscriber's current location and device or group of devices.[2]

The conditional forwarding user experience can be greatly enhanced by combining it with back-end convergence. The resulting combination, while still falling short of full-scale FMC (a subscriber still needs to use multiple devices to connect through different networks, with the exception, perhaps, of Bluetooth CTP profile–enabled cellular phones[3]), nevertheless addresses more of the convergence requirements than other lightweight FMC solutions.

Table 6.2 summarizes the conditional forwarding–based solution features.

TABLE 6.2   Conditional Forwarding Features

Feature	Supported?	How?
Convenience	Yes	Single bill, single point of contact
Coverage	No	Services as well as core and access stay physically separate
Capacity	No	Services as well as core and access stay physically separate
Cost	Partially	Service provider cost can be improved by forwarding to a lower-cost interface
Home-zone pricing	Yes	Possible through pricing and marketing but with a weak business case
Single bill	Yes	Billing systems convergence
Common distribution	Yes	Common distribution
Converged support	Yes	Common tier-1 support team
Fixed-mobile roaming	Yes	Inbound calls only
Fixed-mobile handover	No	Services as well as core and access stay physically separate
Single number	Yes	Pseudo, through permanent forwarding
Converged multinumber	No	Services as well as core and access stay physically separate

---

[2] Similar to cellular systems, which route calls to the serving MSC the subscriber camped on.

[3] Which essentially turns them into cordless devices when they come within range of an appropriately equipped cordless base.

TABLE 6.2   *(Continued)*

Feature	Supported?	How?
Multimode device	No	Services as well as core and access stay physically separate
Converged VM	Yes	Through a common converged VM system
Converged contact list	Yes	Requires an advanced VoIP phone
Find me/follow me	Yes	Possible in VoIP service and certain cellular services
Conditional forwarding	Yes	Needs special equipment in core and customer premises

Having established the viability of conditional forwarding, which does look good on paper, we must admit that despite this class of FMC service represents a significant improvement over back-end convergence, it is yet to be offered by operators on a meaningful scale either by itself or in combination with back-end convergence. Nevertheless, the technology necessary to make this service possible is already available, and we think that this class of FMC solutions has a future, especially as creative new devices and services are introduced to the marketplace.

Use Case: Home Conditional Forwarding

Hans Johansen eschewed a bachelor lifestyle some years ago and is now living comfortably with his wife and two young children in the suburbs. After learning about a new service, "Never Miss a Call," advertised by ConvergeCom, Hans has decided to sign up. Hans' signal in his new home is very weak, so his business partners are often unable to reach him on his mobile number and Hans is too busy for using forwarding which requires engaging and disengaging through *72/*73. He also tried using the *simultaneous ringing* feature but had troubles calling home. Also, his wife often leaves her cell phone in the car when she comes home, so he has to try all numbers before reaching her.

The "Never Miss a Call" service promises to eliminate this inconvenience and provide a few other attractive features along the way. The only requirement of this service is for the subscriber to have a mobile phone with Bluetooth support. After signing up online, Hans receives a package with a small device called CF (short for *conditional forwarder*) and a manual. The device has both the PSTN and Ethernet ports, so it can be plugged into a phone outlet or a router connected to the Internet.

To start using the service, Hans plugs in the device and pairs it up with his mobiles. The service is now operational. Whenever Hans or his wife comes home, their respective mobile phone binds with the CF, which signals the operator's infrastructure to forward all the calls inbound to the mobile number to the landline number so that no calls are ever lost due to a weak signal or a misplaced mobile phone.

(continues)

As before, Hans can access his voice mail and contact list through both his mobile and VoIP phones and can place calls on both phones (providing there is coverage for mobile calls). Hans likes the service. He even decides to install one more CF in the garage so their presence in the home zone can be detected if their mobiles are left in the car.

The service, however, is not perfect: The service only works reliably if Hans's phones are on at the time they enter home zone. Also, Hans and his wife still have to switch among different devices and use multiple separate services (albeit on one account with a single bill and support) with separate numbers. If only the Johansens could get a truly converged service. . . .

Full-Scale FMC

The full-scale FMC user experience builds on many aspects of the lightweight solutions already described and improves on them, providing users with truly seamless connectivity. In the discussion to follow, we are going to use numbering and mode terminology to classify the experience enabled by the full-scale FMC technologies like *dual-* (or multi-) mode telephony or femtocells. We encourage you to refer to the preceding chapter to match the specific scenarios or services we are going to describe with the specific technology making them possible.

3GPP Scenarios   Full-scale FMC is also addressed in several 3GPP scenarios.

The *service continuity* scenario allows the communications service to survive a change of access (technically know as macromobility) between, for example, Wi-Fi and 3GPP-defined systems. The change of access may be noticeable to the user, but there will be no need to reauthenticate the equipment. However, there may be a change in service quality as a consequence of the transition between systems due to the varying capabilities and characteristics of the access technologies and their associated frameworks. It is also likely that some services or functions are not supported due to the possible distinctive treatment of different types of access networks by operators or due to differences in access-technology capabilities.

The *seamless services* scenario goes a step further, specifying *seamless* service continuity or active call handover between different access technologies. Such functionality may be enabled by VCC or numerous proprietary mechanisms recently introduced by the industry.

Finally, the *access to 3GPP CS services* scenario becomes more specific by describing access to services provided by the entities of the 3GPP circuit-switched core network over Wi-Fi. The GAN-based, dual-mode and femtocell solutions provide examples of the technology enabling this scenario.

Classification   The full-scale FMC can be classified in a few different ways. In keeping with the objectives of this chapter, we will classify it according to the user experience and its enablers, such as types and quantities of handsets, number of lines, and perhaps types of underlying infrastructure.

Generally the dual-mode handsets can be classified into the four main categories as follows (note that the behavior description to follow is applicable to mobile phones which are switched on):

- **Single Number/Single Handset (SN/SH)**   Subscribers using solutions falling under this category only need a single multimode handset with a single number capable of communicating over multiple networks such as Wi-Fi, WiMax, and cellular.

- **Single Number/Multi-Handset (SN/MH)**   This FMC solution category allows the subscriber to use multiple handsets (fixed and mobile—akin to today's fixed multi-handset cordless telephony model albeit, extended beyond home zone) to place and receive calls after provisioning a single account. The subscribers are reachable via the same number regardless of their device, location, or connectivity method. Combinations of this service with techniques such as conditional forwarding or simultaneous ringing ensure an optimal user experience. Unified voice mail, contacts, and billing round out the picture.

- **Multi-Number/Single Handset (MN/SH)**   This category allows differentiated conditional ringing behavior. The users carrying one device are reachable via multiple numbers, depending on their preferences, location, time of day, and other parameters. The handset, of course, is multimode.

- **Multi-Number/Multi-Handset (MN/MH)**   This category offers the most complete user experience, allowing many combinations of experiences in multiuser, multiaccess family plans. The subscribers can customize their converged communications systems to their calling preferences and patterns through opting in and out of joining the group (see the description that follows).

Table 6.3 shows the benefits of this FMC solution type.

Single Number/Single Handset   Quite simply, the SN/SH FMC solution enables the same experience as current mobile cellular models. The only distinction is that in the home zone, where wide area cellular coverage may not be available or fixed VoIP service may be more convenient for the operator, or in the hotspots where Wi-Fi coverage is available, calls can be placed and received over a network alternate to the one used for wide area mobile services. Technically this can be accomplished via the use of either dual-mode solutions, femtocells, or any other available technique.

Dual-mode phones in SN/SH may support cellular and Wi-Fi access methods and can be used in wide or local area networks for both voice and data communications. When femtocells are used, the user experience is essentially the same. The only difference to the subscriber is that no dual-mode devices are required; albeit the high-speed data capability offered by Wi-Fi is not available.

Single-number solutions support a single identity for the subscriber in the primary network as well as in the cellular network, which requires coordination between the two networks in intelligent routing of the calls. Customers with a dual-mode handset can seamlessly transition from the cellular network to their home or business Wi-Fi and back.

TABLE 6.3   Full-Scale FMC Features

Feature	Supported?	How?
Convenience	Yes	Single bill, single point of contact, *single service provider*
Coverage	Yes	Flexible common core combined with multiple types of access specific to the environment
Capacity	Yes	Flexible common core combined with multiple types of access specific to the environment
Cost	Yes	Flexible common core combined with multiple types of access, allowing for selecting lowest-cost service options
Home-zone pricing	Yes	Flexible common core combined with multiple types of access specific to the environment
Single bill	Yes	Billing systems convergence
Common distribution	Yes	Common distribution
Converged support	Yes	Common tier-1 support team
Fixed-mobile roaming	Yes	Inter-access roaming through technologies like VCC and femtocells
Fixed-mobile handover	Yes	Seamless handover possible through multiple types of solutions such as VCC
Single number	Yes	Possibly driven by the subscriber needs and specific implementation
Converged multinumber	Yes	In family plans with a flexible common core combined with multiple types of access specific to the environment
Multimode device	Yes	In multimode solutions such as Wi-Fi/cellular
Converged VM	Yes	Through a common converged VM system
Converged contact list	Yes	Requires an advanced VoIP phone

In general, all voice and data services that are available to them in the wide area should also be available in the home zone when connected via the Wi-Fi network through broadband access to the common core.

The solutions in this category are especially well suited for the subscribers simply trying to address poor coverage in their home zone or for others looking to cut cost or consolidate all of their communications needs to a single device reachable over a single number both inside and outside of the home zone. These solutions also offer subscribers and operators cost advantages allowing offload of cellular minutes to less expensive fixed broadband networks.

While SN/SH solutions may not be well suited for most residential family (multiline) plans,[4] they support main target use cases for enterprise and *single-line*

[4] It will be possible to provide service supporting multiple independent sub-accounts (as in today's cellular model), but this category does not allow for a common shared fixed line, which must be purchased separately.

subscriber residential plans (for example, single urban professionals). Being the least complex, they are especially well suited for initial FMC rollouts and early deployments.

The dual-mode handsets purchased for use with SN/SH plans must be provisioned for both cellular and Wi-Fi calling prior to use, which might involve additional steps from the user. Also, in most cases the user will need to install additional equipment such as femtocells when regular phones are used or Wi-Fi routers in home zone for the use with dual-mode devices.

Single Number/Multi-Handset   This category improves over SN/SH in many aspects. The user is no longer limited to a single device in the home zone but can deploy additional handsets such as fixed phones (PSTN cordless phones, VoIP phones, and Wi-Fi phones), all accessible via the same access identifier, which significantly improves usability of the service. The dual-mode Wi-Fi/cellular, femtocell, and other options can be used to support these solutions.

Many calling scenarios are possible in SN/MH. Figure 6.3 presents an example architecture for such solutions in which the interworking gateway, such as VCC,

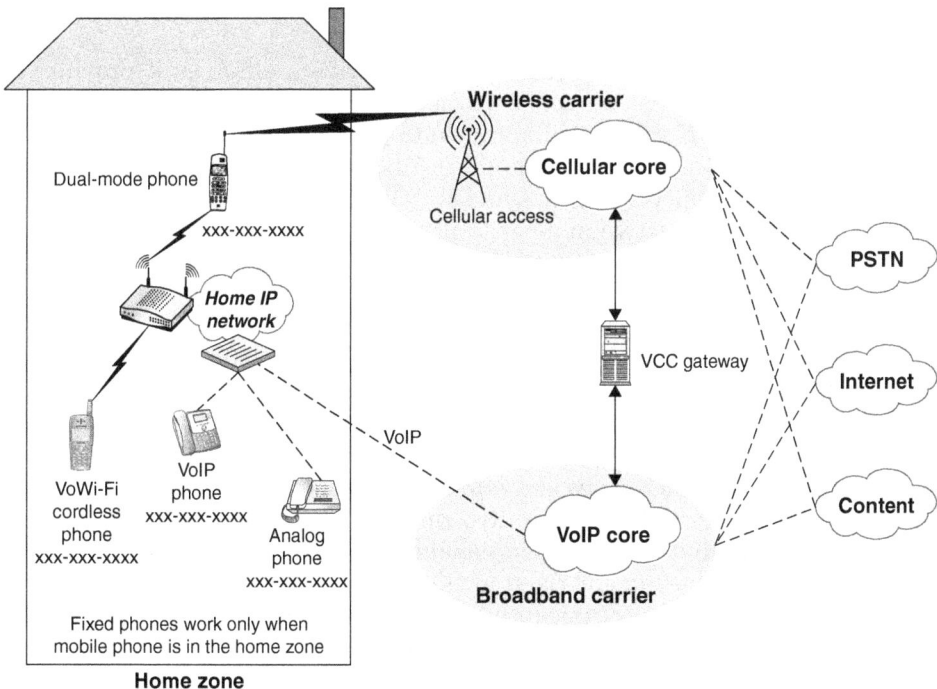

Figure 6.3   SN/MH calling

is used to converge multiple types of access (VoWi-Fi/broadband and cellular in this example). SN/MH solutions go hand in hand with restricted home-zone calling. For instance, the service can be provisioned by an operator to allow calling to and from the home zone only when the subscriber's dual-mode phone is present there. That is, when the dual-mode mobile phone is on and is in the home zone, it will ring on incoming calls together with all fixed phones. Moreover, any of the handsets in the home zone can join the call in progress just like in today's multi-handset cordless solutions. When the subscriber carrying a dual-mode phone leaves the home zone, "the service leaves with him." In other words, the fixed phones are disconnected from the service and cannot be used to place.

Further, the solutions in this category may be configured to prioritize cellular calling over Wi-Fi (use Wi-Fi only when you must) and vice versa (use cellular only when there is no Wi-Fi) or to enforce a one-to-one relationship between a mobile phone and one and only one Wi-Fi AP installed in a particular home zone. When Wi-Fi access is prioritized, the mobile phone would connect to any available (free or appropriately provisioned) access point in the range whenever calls need to be placed or received. The solutions in this category are also well suited for business use, for example, allowing to use multiple types of equipment in the office environment while being reachable through a single number.

Multi-Number/Single Handset   The solutions in this category allow subscribers to share and effectively manage multiple identities on their mobile and fixed handsets capable of supporting these solutions.

MN/SH is useful for operators and subscribers alike, as it provides some unique functionality that may be applied to solve very specific user needs. The target market served by MN/SH solutions may include the customers with the need to distinguish between different types of incoming calls, which, in turn, enables many subscriber identity management scenarios and options. For example, the subscribers may elect to dedicate each respective number to personal or business calling, which together with distinctive ringing and dedicated VM should help them fine-tune their communications to particular situations.

Multi-Number/Multi-Handset   Up until recently, single-number, dual-mode handsets with accompanying service were the Holy Grail of FMC—that is, until, during the various FMC deployments and trials, it became clear that users highly value the ability of a new service to *blend in* seamlessly with their existing equipment such as fixed cordless and corded phones and familiar use patterns.

Simple single- (and even multi-) number solutions proved utterly inadequate in multi-line family account environments where the members of a household may elect to use multiple mobile and fixed lines and multiple types of equipment and may need to be reachable over multiple numbers, depending on their presence and location, time of day, type of call, and other unique conditions.

Apparently other solutions better suited for use with family plans typically purchased by households with multiple mobile phones and fixed lines[5] (so the household is always reachable even when all family members with their dual-mode devices leave the home zone) were needed. Such needs can be addressed with MN/MH solutions described next.

Generally the MN/MH solutions share the following properties:

- MN/MH-based service plans include both fixed (to be used, for example, when no mobile phones are present in the home zone) and mobile lines.

- Fixed phones can be used to place and receive calls at any time whether or not any of the dual-mode phones is present in the home zone. This is in contrast to certain single-number solutions where fixed phones are only usable when at least one dual-mode phone is present in the home zone.

- When in the home zone, dual-mode mobiles either become the part of the fixed-line setup, behaving much like another line in a multiline cordless system, or remain independent, depending on the user or operator preferences (à la SN/SH). In other words, a dual-mode mobile in the home zone may be in one of the two states: joined in (as part of the home-zone system similar to multiline cordless) or independent.

- Missed calls (the ones that were not picked up on any device) and received calls are logged in all devices (fixed and joined-in mobile) capable of displaying them.

When a dual-mode phone provisioned to *join in* arrives in the home zone, it will in essence create an additional line in a home-zone system behaving as a typical multiline cordless solution. Thus when a call arrives on a specific joined-in, dual-mode mobile it will ring on all phones in the home zone. If a new call arrives during an active call, it will not ring on a phone in active call but will display call waiting with caller ID while other phones in the home zone will ring.

Generally, all the joined-in, dual-mode mobiles will appear as additional lines available for calling to all advanced devices in the home zone. That is, all fixed phones capable of supporting the advanced display functionality will show all available lines (available lines = all joined-in dual-mode mobile + all fixed lines), while legacy analog phones and basic digital phones will continue to function in a "single-line" mode. Therefore incoming calls will ring on all "eligible" phones, for example, displaying an "incoming call on line 2" message corresponding with the number dialed by the caller on the other end, as in Figure 6.4. Likewise, one of the available lines can be selected to place outbound calls.

---

[5] This option can also be used by individuals wanting to separate their calling communities (for example, work-related calling and personal calling).

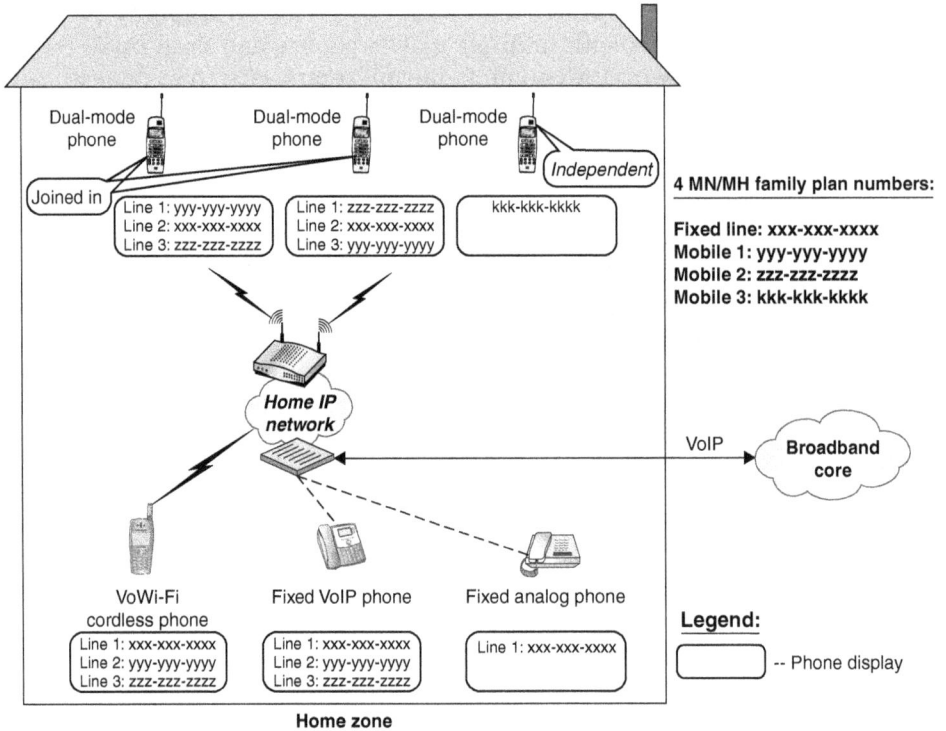

Figure 6.4   MN/MH calling in the home zone

If the MN/MH users elect to keep their phones independent from the other phones in the home zone, their phones will behave as regular mobile phones or SN/SH phones. The only difference (not necessarily known to the user) will be in access type, with calls in the home zone made over Wi-Fi. The following rules will apply in this case:

- A dual-mode phone does not ring on any calls to any other phones in the home zone and cannot join any call active in the home zone.
- No phones in the home zone can join a call in progress on an independent mobile phone.
- A dual-mode phone rings only on the calls inbound to its number and is not aware of active calls on other lines or phones.

A mobile phone in a well-designed MN/MH service should be capable of changing its state (from independent to joined in and vice versa) via some simple user interface (UI) option (for example, selecting the "private" option in the call menu). The subscribers should be allowed to select such a *privacy* option at any time, even during an active call. Selecting "private" in the call menu will change the status of the phone to independent and essentially decouple their mobile phones from the rest of the devices in the home zone.

The operators designing MN/MH solutions must also carefully configure voice-mail system options to provide their subscribers with the service, which takes into account the complexity of the system. For example, the VM system can be provisioned to treat messages destined to a mobile number as personal and therefore deposit them into a private voice-mail box only accessible by a particular mobile. On the other hand, messages deposited into a fixed number VM are also deposited into all voice-mail boxes provisioned for joined-in mobiles and can be retrieved by any joined-in mobile device in and out of the home zone. When the user retrieves the message from any VM box, it should be automatically erased from all other VM boxes.

The operators must also be mindful of a possible race condition that may occur between home answering machines on a fixed line and the network-based VM of a particular mobile phone. That is, if the home answering machine is set up to answer calls after the second ring and the mobile phone VM is set up to answer calls after the fourth ring, the answering machine will be "stealing" the messages destined to the mobile phone. To prevent this situation, the documentation provided to the user has to clearly explain this situation and the steps the user should take to avoid it.

Use Case: Quadruple Play for a Family of Four

Hans' kids are older now, and it is time to upgrade to a new "seamless mobility" family plan, which one of the cable companies in his area has started to advertise. The ConvergeCom service was useful, but it lacked some key features this cable company now offers, so he decides to bite the bullet.

The plan includes four services, as described in Table 6.4: cable TV, broadband Internet access, and a telephony account with one fixed VoIP line and four mobile lines. Hans was already expecting attractive home-zone pricing (unlimited calling from

TABLE 6.4   Quadruple Plan

Services	Equipment	Installation
Broadband Internet access service	Cable modem with router	Plug in power supply and cable from outlet
Fixed telephony service (1 line)	TA adapter in router, two advanced VoIP phones with display	Install cable modem (plug in power and cable, set up Wireless LAN, connect all fixed phones)
Mobile telephony service (4 lines)	Four dual-mode mobiles, Wi-Fi AP, and router	Usual mobile phone one-click provisioning (including both SIM validation and home-zone Wi-Fi security association)
Cable TV service	Set-top box	Install TV set-top box (connect to cable outlet, power up, select auto-provisioning in menu, set clock)

(continues)

home at a fixed, low monthly rate), but he was really after the *seamless* part of the service offered by the advertisement. He also liked the flexibility of the service allowing him to keep all of his existing fixed phones, including the antique black one his grandmother liked, his wife's new cordless phone in the kitchen, and his advanced multiline fixed VoIP office system.

Initially Hans wanted to go "all mobile," but the sales representative suggested adding at least one fixed line for times when no mobiles are present at home and to keep a "household" number (for example, to provide to utility and maintenance people). Shortly after signing up, Hans receives a package with an equipment bundle consisting of the following:

- Four dual-mode (Cellular/Wi-Fi) mobile phones
- One VoIP phone with digital displays and other advanced functionality similar to a typical mobile phone
- A TV set-top box
- A cable modem also supporting Wi-Fi AP, router, and TA functionality

The initial setup was quick out of the box—as in Table 6.4—and involved a few simple steps already familiar to Hans from his previous experience with similar services.

When the Johansens started using the service, they quickly realized how many benefits beyond a healthy cost savings it offers to them. Hans and his wife now never lose calls, since whenever they are at home, both their mobile (that is, when they are on) and fixed phones ring *by default* on all incoming calls no matter what number (fixed or one of the mobiles) is dialed by the calling party. They can also place calls on any mobile or fixed line from any of their mobile phones or one of the fixed VoIP phones (their existing analog phones attached to TA can only be used to place and receive calls on the fixed line[6]). They are delighted to discover that their service and equipment turn into multiline cordless with the number of lines growing as more family members come home.

The service also offers much in terms of flexibility when it comes to privacy. For example, when Hans is expecting an important call from work, he selects the privacy option in the call menu, which effectively decouples his mobile phone from the rest of the home-zone devices so that the incoming calls to his number will only ring on his phone. Hans' daughter, Janet, is a teenager and naturally values her privacy above convenience. Therefore she decided to enable privacy on her phone as a permanent option to make sure no other members of the household can accidentally listen in on her conversations or pick up the calls dialed to her number. . . . *And they lived happily connected ever after.*

---

[6] Restricting calling to Wi-Fi-only when in the home zone is one potential solution that helps operators and users to avoid confusion with regard to home zone vs. wide area calling and potentially minimize support, troubleshooting, and billing issues. No more customer support calls like this one: ". . . I called from home—why am I being charged the regular rate?" The switchover to and from cellular access and Wi-Fi will occur when a Wi-Fi signal of sufficient strength is detected.

Services and Applications Impacting FMC

Undoubtedly the world is getting more connected, with technologies like FMC helping the trend. As a seamlessly mobile environment takes hold, access to services and content is no longer determined by the *availability* of communications media but rather by the subscriber's *willingness and ability* to communicate. As the FMC technology matures and solutions continue to be deployed on a broader scale and gain a loyal following among subscribers, many issues unique to total connectivity will have to be addressed for the FMC service to be commercially successful.

The immediate examples include:

- **Availability** "I want to be reachable only by certain callers, during certain times, or in specific locations I select."

- **Liability** "My customer could not reach me with an urgent issue while I was available and connected via one of the access networks—who is responsible?"

- **Privacy** "I will only share my status with those I allow to monitor it."

The first step in addressing these and other relevant issues is creating reliable mechanisms allowing discovery and use of the real-time subscriber state information, while allowing the subscribers themselves to easily manage their privacy settings.

In fact, the information about the subscriber state is already out there, available in the network and handsets, waiting to be harvested, and with the proper effort and technology in place it can be used to address many potential issues, enable new types of services, and even create a foundation for new applications. The most prominent examples of such information are presence and location, which we are going to explore next in detail.

Presence

Presence is one of those important and useful technologies that to date have largely failed to find their way into mobile networking (in general because of being slightly ahead of its time and for other perfectly valid reasons, such as, unproven business case). However, its success in desktop instant messaging (IM) applications and early forays into enterprise VoIP solutions (e.g., presence-enabled conferencing) proved its general viability, scalability, and usefulness.

Numerous standardization efforts ensued, helping presence to migrate from a concept to a technology. As a result, presence has already become an integral part of some of the mobile technologies such as push-to-talk (PTT)—where it is exposed to the user via active contact lists and used by operators to improve service usability characteristics.

What Is Presence?   Presence is defined as a collection of real-time data describing the subscriber's ability and willingness to communicate across specific media and devices. IETF RFC 2779 [141], dealing with general presence and instant messaging protocol requirements, defines presence service as " . . . a means for finding, retrieving, and subscribing to changes in the presence information (e.g., 'online' or 'offline') of other users."

Generally the presence information in mobile systems can be obtained from three main source types:

- Network elements (registration, location, etc.), or *network agents* in standards terminology

- End-user devices (device status, mode of operation, access network used, etc.) and end users themselves, or *presence user agents* as in 3GPP TS 23.141 (user preferences or user-supplied availability states such as "in the meeting," and so on)

- Elements outside an operator's network (such as subscriber PC or IM clients), referred to as *external agents* by 3GPP

Thereby collected, presence information can then be processed and made available to persons, objects, and processes composing a given communication environment.

Presence Fundamentals   While nowadays in the Internet domain, presence is almost always tied to instant messaging, these two are really independent technologies. IETF RFC 2778 [142], defining the model for a general presence service implementation, further outlines major components of presence technology as follows:

- A *presentity* (presence entity) is an object such as a device, a process, or a person. A presentity may provide presence information such as subscriber willingness and ability to communicate and the way to access presence information.

- *Watchers* or *subscribers* are objects receiving presence information. Watchers may provide information to a presentity necessary to determine their eligibility to receive presence updates from it.

The presence information may be exchanged by presentities and watchers directly or through a centralized presence information repository called a presence server or service as shown in Figure 6.5. Such a repository can be used to aggregate presence from multiple *presence sources* or *triggers* such as subscriber input and preferences, network registration and de-registration, device status, type of access, and operator policies.[7]

At the core of the presence concept are the two operations called *subscription* and *notification*.

Subscription is a mechanism used by a watcher to sign up to receive persistent information about a presentity. The presence information can also be distributed via

---

[7] Note that location can be one of the presence sources. While the subscriber location technology creates a foundation for the entirely new class of applications called location-based services (LBS), in the context of the presence discussion we are going to treat location as just another source of presence information.

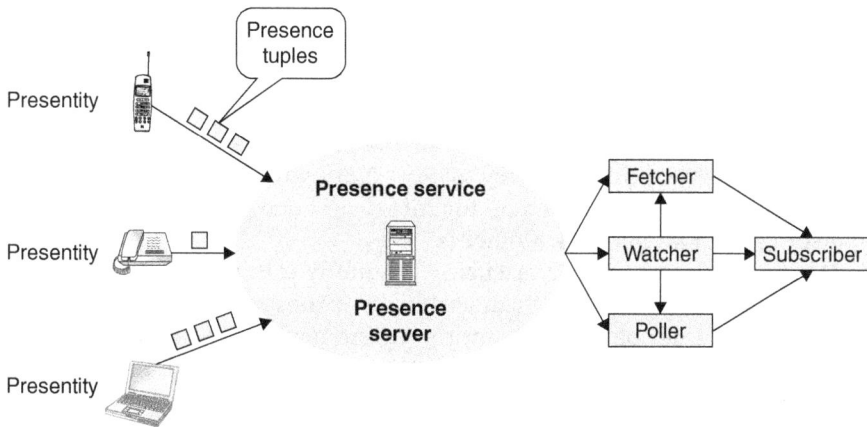

Figure 6.5 Presence service

updates obtained through automatic periodic polling, or on-demand presence inquiries (for example, fetching presence status once per request). Conceptually, all of these methods are similar because they result in the presence status being provided in response to a request by a watcher. That is, a presence server or presentity itself *notifies* a watcher about changes in the subscriber status.

A subscription for presence notifications (arguably one of the most popular ways to obtain presence) provides for a mechanism allowing a watcher to "subscribe" to monitor a presentity's presence status. As a result of such a subscription, the presentity notifies the watcher whenever its status changes without the need for a specific query. Generally these notifications take the form of specific messages defined by a particular protocol.

Another core component of presence technology directly related to subscriber privacy management is *authorization*. During presence transactions, significant amounts of information about subscribers are collected and made available in the network. So the operator must put in place a rigorous process determining who can access this information. For example, the subscriber may wish to restrict the information made available to certain watchers according to their identity, time of day, or other conditions: "notify only my customers of my status change today."

Finally, we must establish the distinction between presence and *availability,* a topic much debated in the industry in recent times. However important the availability might be, in our view, subscriber availability is just another presence variable. The users' *ability* to communicate is not always equal to their *willingness* to communicate; hence a measure is needed to indicate this desire. The availability, therefore, is defined as a measure of a subscriber's willingness to communicate. For example, selecting "in the meeting" status may automatically route all incoming voice calls to VM along with a mere status notification in accordance with a specific preset policy supported by the operator.

Note that presence is a "perishable" product. That is, it represents the information about the user at a given moment in time. The accuracy and, consequently, the value of presence is proportional to frequency of presence updates. Indeed, the lower the sampling rates, the more chances for the users to change their status between measurements. That presence property is also one of the major reasons for its slow adoption by the *mobile* networking industry to date. Naturally, the more frequent are the presence updates, the higher the load on air interfaces, end user devices' processing power and power drain, and network elements.

These barriers, however, are being gradually torn down as the air link bandwidth increases and the cost per bit transmitted over the radio interface lowers, devices become smarter and their battery life improves, and new core technologies—such as IMS, with its native presence support—are deployed. With these changes, the industry is now taking a second look at the presence technology and business case, with FMC solutions being primary candidates to use presence as a parameter for access method selection: "I prefer to be connected through Wi-Fi [if available] when my presence status is set to 'in the office.'" So the presence status information becomes an implicit portion of a converged communication profile.

Standardizing Presence   In part spurred by the emergence of desktop IM applications such as AOL, Yahoo, and MSN, which relied on proprietary presence protocols, numerous efforts have been put in place to standardize presence. These efforts were in part driven by the lack of interoperability between these systems. For various reasons, a single standard never emerged.

The standardization of presence is still evolving and is currently being undertaken by a number of standards organizations, such as IETF, 3GPP, and Open Mobile Alliance (OMA). As a result, as many as five competing presence standards coexist, though one of them, the Session Initiation Protocol for Instant Messaging and Presence Leveraging Extensions (SIMPLE), has recently emerged as a standard of choice for commercial applications in mobile and fixed systems.

Next we provide a quick overview of major presence standards.

XMPP   The Extensible Messaging and Presence Protocol (XMPP), defined in IETF RFC 3920 [143], is an XML-based presence and instant message exchange protocol developed based on the specifications from the open-source Jabber[8] organization. The XMPP-based architecture provides gateways to multiple presence and messaging protocols that map into the common XML format, thereby allowing heterogeneous IM systems using different protocols to interoperate.

The main advantage of XMPP lies in its use of the flexible and easy-to-implement XML format. This format, however, is very verbose and is not well suited to the limited capabilities of mobile devices and air interfaces with inadequate bandwidth.

---

[8] Jabber was originally created to unify proprietary IM protocols prior to the creation of the Instant Messaging and Presence Protocol group in IETF.

The open-source nature of the protocol has led to some penetration in the desktop market, but it has not seen much commercial adoption, especially in mobile wireless networks.

SIMPLE   An Instant Messaging and Presence Protocols (IMPP) working group, created in the IETF in 1998, was chartered naturally with the definition and eventual standardization of the presence and instant messaging protocols.[9]

After several years of work, the working group settled on creating a common abstract format for presence and messaging and allowed for the possibility of multiple protocols consistent with this format to coexist. One of these protocols, called SIMPLE, quickly gained the lead and has now become the dominant presence standard, especially after its adoption by 3GPP and the OMA. Incidentally, SIMPLE also refers to the working group in the IETF that is producing specifications for presence and IM.

SIMPLE is based on SIP, and in addition to acceptance by the main standard bodies, it is now backed by the major industry leaders for both fixed and wireless services and equipment. At the core of SIMPLE is a SIP element called a *location server* that is designed to maintain the addresses of SIP clients. SIP sessions are initiated after the addresses for the SIP endpoints are established from consulting the location servers.

SIMPLE, with its foundation defined in IETF RFCs 2778 [142] and 2779 [141], provides a set of extensions to SIP designed to support presence service requirements. These extensions are

- SUBSCRIBE, defined in IETF RFC 3265 [144] and 3856 [145], allowing a watcher to subscribe to receive presence updates.

- NOTIFY, also defined in IETF RFC 3265 and 3856, allowing a presentity to notify subscribed watchers of its status changes triggered by subscriptions.

- PUBLISH, defined in IETF RFC 3903 [146]. Publication is a "push" of raw presence data into the system by sources of that presence data. As an example, a handset might publish its presence status to presence server whenever it changes.

The presence information in notification messages is carried in capsules called *presence documents*, commonly encoded using XML according to the format outlined by Presence Information Document Format (PIDF) defined in IETF RFC 3863 [147]. PIDF assumes that the presence information of a presentity is broken into a set of *tuples*. A tuple is associated with a user device. For example, if a user has a PC and a phone, that user's presence document would contain two tuples—one indicating the status of his or her PC, and the other indicating the status of his or her phone.

---

[9] Quoting the IMPP charter statement: "This working group will eventually define protocols and data formats necessary to build an Internet-scale end-user presence awareness, notification, and instant messaging system. . . ."

SIMPLE specifications also include the presence *event packages*. Two of the most notable are

- A *registration* or *RegEvent,* defined in the IETF RFC 3580 [148], providing notifications of registration changes
- *Watcher Info,* defined in the IETF RFCs 3857 [149] and 3858 [150], to let clients know who has subscribed to watch their presence status, to manage permissions to watch, and to manage other privacy aspect

Another important property of SIMPLE is its ability to subscribe to and manage *presentity lists* as opposed to individual presentities. In many cases, a potential watcher may need to monitor the presence status of more than one subscriber, such as in the case of a presence-enabled contact list on a phone containing multiple entries. Essentially, *list subscriptions* allow clients to watch multiple presentities with a single subscription. For example, a list subscription feature allows a presentity to subscribe to watch a group of users or multiple contacts as opposed to single users.

Presence in 3GPP   3GPP stage 1 and stage 2 presence services are defined in TS 22.141 [152] and TS 23.141 [154], respectively. These documents—largely based on IETF RFCs—lay down the foundation and framework for a generic presence service in mobile environments and outline a set of requirements for a practical implementation of a presence solution. The presence service architecture defined in these documents encompasses both the Network and Application layers as well as end-to-end presence signaling flows.

The Network layer of the presence service defines the communication between the presence servers and other presence elements and standard 3GPP core network elements such as MSC, HLR, and SMCS. The Application layer of the presence service defines the communication between the presence service elements (e.g., between watchers and presentities).

The stage 2 presence service definition also sets forth presence reference points such as P_{eu}, P_{en}, and others, as in Figure 6.6, defining interfaces to network elements, which can be used as potential sources of presence information (see the following text). The 3GPP TS 23.141 also defines the functionality of a presence server and relationships between a presence server and presence agents. Finally, 3GPP TS 24.141 [153] formally defines SIMPLE as a 3GPP standard for presence and identifies the presence procedures allowing IMS elements such as CSCFs to interact with presence agents and a presence server.

Presence in OMA   The Open Mobile Alliance (OMA) has been engaged in defining presence for some years along with IETF and 3GPP. The activities were primarily conducted by the Instant Messaging and Presence Services (IMPS) workgroup, set up by major handset manufacturers Nokia, Ericsson, and Motorola, to create a set of specifications for propagating instant messages and presence information between a mobile handset and a server in the mobile network as well as among the servers in the network.

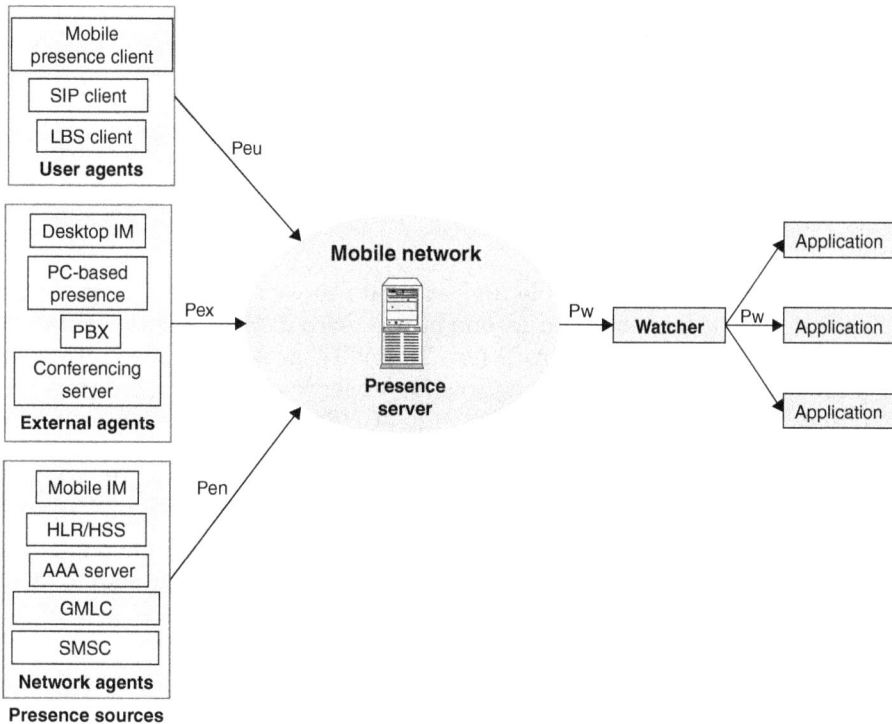

Figure 6.6   3GPP presence architecture

The OMA IMPS set of standards includes the protocol for communicating presence information between servers (the Server-to-Server Protocol, or SSP) and the protocol for communications between the client and the server (the Client-to-Server Protocol, or CSP). SSP is typically used to synchronize presence information from external servers, and for performance reasons, a messaging server will keep its own local presence database, thereby duplicating the presence service.

Unlike other presence standards, the OMA IMPS protocol has been developed from the ground up to support presence in mobile environments, taking into consideration the limitations of mobile devices and networks. On the other hand, the main disadvantage of deploying an OMA IMPS presence service is its lack of extensibility for any applications not supporting this standard. Also, an IMPS presence server cannot be shared by multiple applications, since the structure of presence information is specific to IM and not easily extensible.

In parallel with IMPS activities, OMA also created a set of presence service specifications based on SIMPLE and closely tied to 3GPP work. Primary documents defining OAM SIMPLE-based architecture are the Presence Simple Specification [155] and the Presence SIMPLE Architecture [156].

The OMA SIMPLE specifications add to 3GPP definitions presence interaction between various OMA presence entities on the Application's layer, such as

- Presence information handling via PIDF
- Presence information handling via RPID
- Presence location information provided via the GEOPRIV location object as defined in IETF RFC4119 [151]

Detecting Presence   Reliable and accurate presence information gathering or *presence detection* has emerged as one of the more difficult challenges in the early days of presence technology formation. Typically, presence information is collected by allowing a presence server to aggregate presence states from multiple *presence sources* (or *suppliers of presence information* in 3GPP terminology) to create an accurate real-time representation of the subscriber availability. Thus-gathered information may include registration state, call state, data session state, application state, and the subscriber communications preferences as in Figure 6.6.

The presence server in this figure, acting on behalf of a watcher, may, for example, get pushed periodic presence updates from a particular presence source by using SIMPLE Publish. Alternatively, whenever presentity status changes, it will *notify* the presence server. The presence sources generally fall into three groups:

- Network elements, or network agents in 3GPP terminology
- Subscriber and handset application input, or user agents in 3GPP
- External presence sources, or external agents (that is, the ones not governed by the service provider's network)

The presence information thus gathered by the presence server is continuously processed to create an accurate view of the subscriber state. Table 6.5 lists potential presence sources in each category along with the associated interfaces to the presence server.

An example of a user agent could be the mobile handset supporting a variety of the client- or subscriber-based input or specific policies. Thus an FMC handset with the SIP client may support SIMPLE extensions or simply (no pun intended) register with the SIP location server by sending SIP registration messages.

Other examples may include Mobile IM, calendar client, and similar applications allowing users to manually enter presence status ("in the meeting," "away," "at lunch," etc.) and propagate it to the presence server. Finally, the LBS application, combined with the user- or operator-defined policies, may supply device coordinates to be used to enforce conditional actions ("I am not available for calls from office PBX when I am in the home zone or in Bolton Landing, NY, U.S.A., in my vacation home").

The examples of network agents include common cellular core network elements such as HLR and SMSC or radio interface components such as BSC designed to keep track of end-user equipment status and whereabouts. A presence server theoretically should be able to pull presence information from such network sources through

TABLE 6.5   Examples of Presence Sources (Mobile Operator View)

Source	3GPP Reference Point	Interface	Protocol/Message
Network Agents			
HLR/HSS	Ph	MAP IS-41 IN (CAMEL, (WIN 2)	MAP ATI IS-41 LOCREQ IS-41 POSREQ IS-41 O/T Answer IS-41 O/T Disconnect
AAA server	Pr	IP RADIUS DIAMETER	RADIUS Start/Stop
CSCF	Pi	SIP, IP	Register
MSC/VLR	Pc	X	SMSREQ
GSN/PDSN/HA	Pg/Pk		GTP, Mobile IP
GMLC	Pl		Various
PDG	Pp		Various
IMS VCC	NA	SIP, SIMPLE	IP
User Agents			
SIP client	P_{eu}	SIP	SUBSCRIBE NOTIFY PUBLISH
OMA IMPS client	P_{eu}	CSP/SSP	SSP/SCP
XMPP client	P_{eu}	XML	Proprietary
Client LBS application	P_{eu}	Location Object	Proprietary
Client-based calendar	P_{eu}		MIME, HTML
Mobile IM client	NA	Proprietary, SIMPLE	Proprietary SUBSCRIBE NOTIFY PUBLISH
External Agents			
Desktop IM client	P_{ex}	SIP, proprietary	SUBSCRIBE NOTIFY PUBLISH
PC-based presence client	P_{ex}	SIP, SIMPLE	SUBSCRIBE NOTIFY PUBLISH
PC-based calendar	P_{ex}		MIME, HTML
Enterprise PBX	P_{ex}		IP, SS7, various
External Presence server	P_{ex}	SIP, SIMPLE	SUBSCRIBE NOTIFY PUBLISH
Conferencing server	Various	Proprietary	IP, proprietary

periodic polling, triggering, and other mechanisms. While this functionality is not fully defined in 3GPP, a number of proprietary solutions have been designed with the ability to discover presence from new sources within a network.

However, what looks good in theory does not always work in practice. Mining networks for presence often proved to be difficult, impractical, or even impossible. Let's for example consider discovering presence from the circuit cellular core using Intelligent Network (IN) triggers. The carrier's HLR and VLR can be a treasure trove of subscriber data such as MS status and location. To get access to the information about a specific MIN or MSID, the presence server must *interrogate* the HLR via one of the IS-41 or GSM MAP functions, such as LOCREQ or CAMEL Any-Time-Interrogating (ATI), respectively.

According to the operator preferences, the HLRs can be configured to query (page) the mobile every time it receives the LOCREQ request from the presence server appearing to it as an MSC. HLR can also be triggered to release MIN information via SMSREQ messages or O/T Answer and Disconnect. Figure 6.7 depicts a sample HLR interrogation state diagram.

Despite its apparent benefits (no need to query handsets directly, thereby wasting valuable radio interface bandwidth and airtime), this class of presence detection methods never gained much popularity with carriers. First, this approach generates high loads in SS7 networks, where capacity is always at a premium. Second, it is relatively inaccurate when HLR is only triggered for subscriber state changes because certain HLRs are provisioned with fairly long inactivity timers, sometimes requiring hours to detect the handset OFF status (in case of "non-graceful" sign off, for example, as a result of breakage). Finally, the option in which HLRs were set to interrogate the mobile whenever presence status was requested negated the radio interface efficiencies promised by this method.

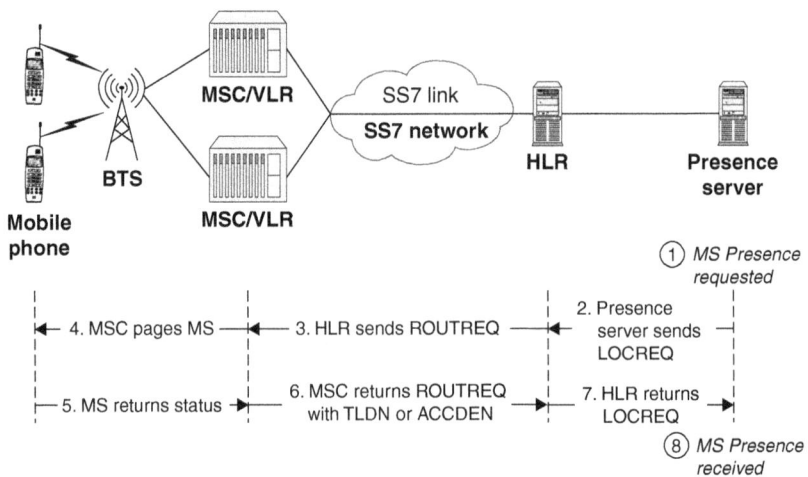

Figure 6.7   HLR interrogation

LBS

Location-based services (LBS) provided in a cellular (or other wireless) domain deliver value to the subscriber or third party based on the ability to determine the subscriber's device location in real time. The information about the device location can be used by the applications running on the device itself or by the systems and process based in the network. Further, the device (and most often presumed subscriber) location information can be passed on to the external entities or shared with the subscriber peers.

In its relatively short (less than a decade) history, LBS initially enjoyed great expectations followed by market hype, only to see a gradual loss of interest from both consumers and service providers. The immaturity of the applications and end-user equipment, coupled with early pains in defining an increasingly complex value chain and application ecosystem, soon ended all hopes to realize LBS market potentials.

This situation, however, was turned around in recent years by concerted industry efforts and successful introduction of new business models targeting specific subscriber segments with a few well-thought-out applications and services designed to address immediate subscriber needs such as turn-by-turn navigation, people and asset tracking, and point-of-interest (PoI) guidance. Both business and residential consumers nowadays are embracing mobile LBS despite the increasing competition from other types of location devices and services such as stand-alone GPS-based navigation systems, as they have rapidly fallen in price.[10]

The importance of LBS for FMC solutions is hard to overestimate. Besides the obvious ability to select the access network and other parameters of service based on the subscriber's location, LBS and FMC solutions provide other ways to enhance their mutual value propositions.

For example, a seamless multiaccess environment enabled by FMC improves LBS accuracy and offers the capability to track subscriber locations beyond the cellular (or even GPS satellite) coverage area. On the other hand, FMC can use LBS to provide subscribers with more options in access selection and new finely tailored services, such as location/access pair-specific content delivery and differentiated billing as well as "virtual home zone" solutions.

LBS Technology Primer   The key component to any location-based service is a positioning system based on *location servers* called Position Determination Entities (PDEs) in 3GPP2-defined core networks or Mobile Location Centers (MLCs) in 3GPP environments. These platforms augment the Global Positioning System (GPS)[11] data received

---

[10] It is important to note that off-the-shelf navigation systems either mounted in vehicles or offered to the consumers in the form of portable devices do not support the types of LBS services possible in cellular environments.

[11] A 29–*Navstar* satellite constellation funded and built by the U.S. government to enable free high-accuracy positioning based on satellite trilateration.

by the device with the information derived from cellular network base station *trilat-eration*[12] and at the same time disseminate the geographical coordinates (latitude and longitude) of the device to other systems, processes, and devices in the network. Sometimes, also, the device may not be GPS capable or the GPS signal may not be receivable due to obstacles, so the network-based solution becomes essential.

When an application needs to determine handset location, it sends a location request to a location server (PDE or GMLC) via the LBS middleware. The location server then determines the device location by interacting with one of the positioning systems described in the next sections and provides the information to the application.

Cell-ID   Cell sector identification, or Cell-ID, is one of the earliest and most basic wireless location technologies. This technology does not require GPS, relying instead on a serving cell to provide subscriber location information with relatively low accuracy (hundreds of meters). When a mobile phone connects to the nearest base station, it transmits its own identification number (Cell-ID), the identity of the phone making the call, and the sector from which the base station antennas are picking up the signal.

Since base stations are typically situated approximately a kilometer apart, the area defined by the cell-sector ID can be measured in square kilometers. The closer base stations are to each other, the more accurate the mobile location determination.

E-OTD/U-TDOA   Enhanced Observed Time Difference (E-OTD) and Uplink Time Difference of Arrival (U-TDOA) technologies go one step further than Cell-ID by trilaterating the mobile location by measuring time differences of signals arriving from *more than one* base station. This technique allows for dramatic improvements in accuracy over Cell-ID. While U-TDOA is used in both GSM and CDMA networks, the E-OTD solutions (less reliant to synchronization native to CDMA networks) are mainly deployed in GSM.

In both techniques, handsets are located by measuring the timing of the signal arrival from the handset to multiple base stations it is attached to. The equipment performing such measurements and sending the information to the MLC or PDE is called a location measurement unit (LMU). Both solutions allow operators to achieve positioning accuracy of 50 meters or better—the denser the base stations, the higher the accuracy, just as with Cell-ID.

A-GPS   Assisted GPS (A-GPS) combines GPS locationing with base station trilateration to provide an accurate mobile location when either GPS satellite or cellular coverage is available. The word *assisted* in the name of this system signifies that a location server is used to assist a mobile to obtain its location. A-GPS was designed for use in

---

[12] Trilateration is defined as the determination of a position based on the distance relative to other known location(s).

CDMA networks and is primarily used in the regions where CDMA is deployed, such as North America and Asia. A-GPS supports two primary modes of operation:

- **MS-based**   The mobile device calculates location based on the GPS signal
- **MS-assisted**   The mobile device relies on the location server to calculate location

AGPS has significant advantages over both stand-alone GPS and cellular-only trilateration methods, incorporating their best features to achieve highly accurate performance.

First, it improves time to first fix (initial location determination) and sensitivity, in most cases bringing it to less than fifteen seconds from a cold start (initial startup). Second, it provides higher positioning accuracy (in most cases under 10 meters). Further, it provides the ability to calculate mobile locations even when either cellular coverage or a GPS signal is not available. Finally, AGPS can be potentially extended to allow a location server to interact with noncellular access networks like Wi-Fi or WiMax (making it especially applicable in an FMC environment).

LBS Classification   The solutions based on LBS can be broadly divided into two categories: *device* and *network* based, as depicted in Figure 6.8.

Device Based   Device-based LBS applications are launched by the subscribers from their mobile devices either for use by the subscribers themselves or to provide other humans or machines with their location information. This class of applications

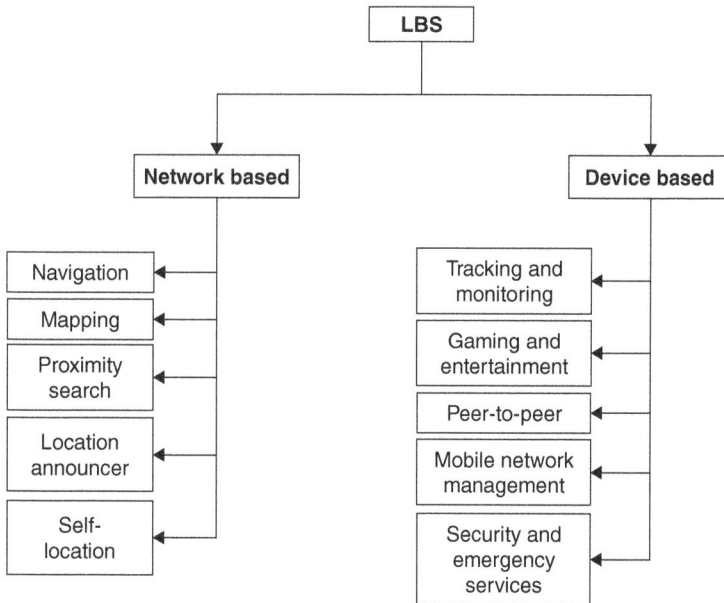

Figure 6.8   LBS classification

generally requires appropriate clients to be supported in the mobile device along with the prerequisite location-determining capability. It also requires device interaction with the network both for determining its geographic coordinates and guidance directions, mapping, and PoI data.

Some of this information can also be stored locally in the device, thereby reducing service dependency on network connectivity in certain cases (for example, remote areas with poor coverage). The most popular examples of the applications in this group include

- Personal self-location (e.g., displaying the subscriber's location on the map resident or downloaded to his or her devices—"Where am I?")
- Personal navigation (e.g., static or real-time "turn-by-turn" guidance directions to a certain address or PoI)
- General mapping (e.g., "follow me" continuous map display or traffic map display along the route or around the current location)
- Proximity search (e.g., nearby PoI search or browsing)
- Location announcer (e.g., an application supplying a subscriber location to "friend finders" and other watchers [borrowing from presence terminology])

Network Based   Network-based applications, which may or may not reside in the service provider's network, typically rely on the information about the device's whereabouts to perform actions, which generally involve the device but also may be used for other purposes. Many of the services in this category can provide significant benefits to FMC operators and subscribers. Typical examples of the applications under this category include

- **Device (subscriber) tracking and monitoring**   The services under this category can be used to obtain locations of individuals (children, field force, seniors, etc.) or equipment and machinery, such as vehicles and agricultural and construction equipment (most such tracking services benefit greatly from the indoor coverage offered by FMC, which offers the possibility to obtain location information from other networks, such as Wi-Fi).
- **Location-based gaming and entertainment**   The applications in this group allow for gaming, content sharing, and other types of entertainment based on location ("I will play Doom with anyone in Nacogdoches, TX, U.S.A.").
- **Peer-to-peer**   The applications in this group, for example, *friend finders* and *content sharers,* which are usually hosted by network-based servers, allow the subscribers to get in touch with each other in terms of their geographical locations ("Inviting all Doom Players Club members within 10 Km from my location to play a quick one").
- **Location-based commerce**   This group may include applications like targeted advertising when, e.g., promotions or discount coupon offerings to subscribers are triggered by their immediate proximity to a retail location (more on this in Chapter 7).

- **Mobile network management**   Here the device location is used by the network to make routing and RF planning decisions and network access and content-throttling selections when, e.g., deployed in conjunction with FMC solutions. Femtocell location tracking also falls under this category.

- **Security and emergency services**   Known as E911 in the U.S., these services are often mandated by regulators to allow emergency and law enforcement agencies to quickly determine the location of citizens requiring assistance or in distress. The support of these services requires installation of positioning systems in the mobile networks. Certain other services such as roadside assistance also belong under this category.

- **Location-based billing**   Earlier examples of home-zone pricing are perfect illustrations of the usefulness of this technology, which allows for differentiated billing based not only on the time of day and access method but on location, such as the home zone, greatly enhancing the FMC value proposition.

Summary

In this chapter, we took a bit of a detour from the discussion of FMC business and technology and concentrated instead on the user-experience aspects for FMC, providing plenty of use-case scenarios covering the most promising approaches. We also classified the FMC solutions in terms of handsets and access numbers, taking a purely user-centric approach, since focusing on the end-user needs and drivers is especially important for the complex technologies like FMC undergoing the formation stage. We have finally explored the aspect of presence and location-based services, as an enabler of advanced converged applications.

We are now ready to wrap up the discussion with the final chapter of this book, attempting to look into the future of FMC and mobility as a whole.

Looking Ahead

Anything that is theoretically possible will be achieved in practice, no matter what the technical difficulties, if it is desired greatly enough.

—Arthur Clarke, *Profiles of the Future*

This long journey through the interesting, albeit sometimes complex, FMC world has so far taken us through business scenarios analysis, technology overview and assessments, and real-world deployment strategies. This "voyage of discovery" would be really incomplete if we did not consider sufficiently the future directions this technology may take and its impact on the end-user lifestyle and/or day-to-day business. While doing so, we are not going to attempt to predict the future. Rather, we will extrapolate the current industry trends and extend FMC's current voice services focus to other technologies, thus trying to outline a medium-term scenario for FMC and, more generally, an evolutionary path for telecommunications as a whole.

As always, the level of public acceptance of FMC will be the ultimate judge of its long-term viability and commercial success. The uptake levels of this novel approach to communications will either encourage operators to quickly move away from early trials and limited rollouts into mass-scale deployments, or force them to reconsider the directions of their early investments.

Meaning of Mobility Revisited

The telecommunication service experience so far has been defined by a taxonomy identifying three major paradigms:

- Fixed
- Nomadic
- Mobile

In keeping with the definition of "fixed" from the first chapter, the users of *fixed* telecommunication services do not experience mobility at all, and the typical devices used for such services include, for instance, a classic tethered PSTN telephone or a cordless telephone.[1] Within this category also fall narrowband and broadband Internet access services and, correspondingly, fixed VoIP telephony. The important common characteristic for these services and devices is that they can be used only in a single location permanently defined by an operator and provided by fixed facilities such as outside-plant copper or fiber termination points in the user's home or office.

Nomadic services are an extension of fixed-line services, as they are used in the same way from a mobility standpoint. That is, while the nomadic service is available in multiple locations, the user is not allowed to move away for a meaningfully long distance from the point where the service had started (not unlike with the fixed services). Moreover, the user cannot automatically switch between different locations and preserve (i.e., keep alive) an active voice call or data session without the need to reconnect and often reauthenticate. Therefore if the user must roam between locations, the resulting services will be fragmented and will have to consist of a succession of disjointed voice or data sessions, which will have to be manually reestablished by the user.

Typically, *nomadic* services are implemented by allowing users to enter through public access points, upon authentication (identity validation) and authorization (e.g., account status check and access rights confirmation). These services could be paid fixed public telephones[2] accessed with calling cards, or a network of Wi-Fi hotspots allowing access to the Internet. Such services could be offered, for instance, based on explicit direct subscription to nomadic telecommunication services (e.g., the user takes out a calling card service, which is intended for use while on the move) or as an extension of fixed or mobile telecommunication services (e.g., access to Wi-Fi hotspots could be bundled either with a fixed-line Internet access contract or with the subscription to mobile phone service so that users can get connectivity outside their home zones).

Furthering *nomadicity, mobile* telecommunication services, first defined in the beginning of the book, allow users to maintain uninterrupted connectivity while changing their physical location, with the typical example being cellular wireless services.

Old Taxonomy Obsolescence

The taxonomy described in the preceding section, however, is destined to be made obsolete by the introduction of FMC. Its underlying assumptions that the user is bound to access telecommunication services via a single type of access at a time, and that the subscription is bound to an access technology or method, are no longer valid in the converged environment.

---

[1] Technically cordless telephones allow a certain degree of localized mobility, but so does a tethered telephone with a sufficiently long cord.

[2] Fast becoming part of the past.

A single device and a single subscription can bundle together a set of access methods and non-homogeneous services, so that there is no longer a distinction, from a user's standpoint, between a service used in the home zone (e.g., in the coverage of the Wi-Fi access point or femtocell) or in an airport hotspot, or while enjoying a day on the beach using a wide-area cellular network! Indeed, the same service can be provisioned and used as fixed, nomadic, or mobile depending on specific subscriber needs at any point in time.

Bundling and Unbundling   Not only this, with FMC it is also possible to bundle together a set of *access network* subscriptions and *service* subscriptions. For instance, the same services bundle can be accessed by a user via Wi-Fi connectivity or wide area cellular connectivity when Wi-Fi coverage is not available. Wi-Fi hotspot access may be granted by using the home-zone broadband access subscription, which allows global Wi-Fi hotspot roaming.

Likewise, wide area cellular subscriptions can be obtained from a wireless service provider. Such a subscription bundled with a set of both services and access technologies can be provided by either fixed or mobile operators, or by *converged operators*.[3] A fixed or mobile operator would have to arrange alliances in the "home country"[4] to enable such bundled offering. A converged operator would be able to offer all bundles without setting up alliances with other access operators.

Expanding this logic further, it is easy to see how FMC enables *unbundling* of access and services when there is a need to position only converged access or access-independent service offerings (for example, seamless video streaming over any broadband access available) with particular subscriber segments or to address the needs of the population or businesses in specific geographies.

Removing the Barriers   These FMC properties, allowing for easy bundling and unbundling of services and access mechanisms, essentially challenge the very definitions of *fixed* and *nomadic* services, making them indistinguishable from *mobile*. Under FMC, a user can experience the same services, regardless of the location, time, and mobility pattern. Using one access type or another no longer matters, and in fact users may not notice changes between them.

In fact, the level of awareness about the differences in using a service over a particular access method will probably be a function of different tariffs or a service adaptation applied when a user is in a home zone or in a wide area, in the office or on the road. For instance, while in the home zone, an FMC terminal may ring also when the customer is called on the fixed-line number, while this may not happen when he or she is on the move. In addition, the price of a call placed while in the office may be lower than that of a call placed while on the road.

---

[3] By converged operator, we mean an operator that owns both fixed and mobile accesses outright and can therefore combine these in access service bundles without partnering with other providers in the home country.

[4] We define *home country* to be the territory of operation where the subscriber is considered nonroaming.

To ensure a satisfactory user experience, the subscriber may, for instance, be advised about the specific access method being used and associated tariffs or other conditions. Such "advice of charge" can be delivered through documentation or specific user interface messages in the subscriber's device (e.g., a home-zone logo on the screen, a "leaving home zone" notification as an audible tone, or a "high-speed network available" message).

In summary, with the introduction of FMC, the barriers between mobile, nomadic, and fixed communications are removed, as they blend into a seamless telecommunications environment and benefit from their respective services, by making them available ubiquitously and helping to make mobility pervasive. The following sections discuss what's possible with convergence as well as what challenges are likely to be encountered along the way.

Driving Toward Uniform Connectivity

The blurred boundary between fixed and mobile and between wireline and wireless that characterizes FMC solutions and services is a distinctive feature of its value proposition and one of the primary novelties introduced in telecommunications in recent years. However, its novelty and even its value alone would not really guarantee broad adoption and market acceptance if an ecosystem of services targeted at this new landscape would not emerge.[5]

New Application Ecosystem

A new way to approach the delivery and implementation of services is essential for the creation of the FMC applications ecosystem. This objective can be achieved with a new service delivery platform, which would provide application developers with unique features enabling them to create exciting new services. Such features may include always-on IP connectivity with a constant, access-independent IP address, mechanisms to implement policies or preferences in access technology selection or preferred points of contact at different times of day, and continuation of voice calls in the face of technology changes.

How can applications leverage the vast array of possibilities, which can be offered by an FMC service delivery platform? The answer to this question is not so easy. Also, the way such applications will be vertically integrated with the underlying networks is not at all predictable, albeit at the beginning it can be easily foreseen that the owners of the access networks will be intimately involved in supporting both the FMC service delivery platform and the majority of its applications.

Vertical Integration   Vertical integration, a necessary element of new application ecosystem, benefits from tighter coupling between accesses and services. This allows faster time to market, as the openness and standard interfaces needed to run certain FMC applications

---

[5] Geoffrey A. Moore outlined this need in his book *Crossing the Chasm: Marketing and Selling High-Tech Products to Mainstream Customers* (Collins, revised edition 2006), which analyzes how technologies cross the chasm between the early adopters and mainstream acceptance.

may not be necessary with proprietary, vertically integrated applications. Vertical integration would therefore allow differentiation *before* standards-based solutions are available.

On the other hand, vertical integration would limit the application space, as the necessary ecosystem of applications developers would simply not exist or would be proprietary in nature, which in turn would not allow application vendors to realize any economies of scale and would make it more difficult for operators to source applications from an ecosystem of competing application suppliers.

Service Delivery Ecosystem Characteristics   What will the main characteristics of the FMC service delivery ecosystem be? It can be expected that it will be possible to gather real-time information on users' status, including presence, reachability, and location (as described in Chapter 6), as well as access technologies' availability and parameters, like signal strength, for purposes such as accounting, technology selection, and uses yet to be foreseen. It will also be necessary to synchronize service settings and status across independent domains (if these are not using a common centralized back-end system to run the service logic). User preferences and applications-related user data may be stored both in the terminal and in subscriber data repositories.

The information stored in these centralized subscriber data repositories, capable of storing the user profile and context information, in turn could be retrieved by different access networks. However, centralized service logic, service settings, and subscriber service profiles may be utilized only within a single customer-provider relationship. If it is admissible for an FMC subscriber to use services from different providers, it then must be assumed that multiple subscriber information databases will coexist and multiple services execution environments will be concurrently available to a user.

To support such a "centralized/distributed" model, reaching beyond the current IMS vision, it will be necessary to create a method to gather, reconcile, and exchange the information governed by different subscriptions. It can then be foreseen that some "user data" peering mechanisms, similar to current voice and GPRS Roaming Exchanges (GRXs) and VoIP ENUM services, will be required among the players in the FMC space. Therefore, the prevalent degree of coupling (i.e., of vertical integration) and information sharing between the access operators and service FMC back-end systems will be dependent not only on the technology but also on creation of the exchange ecosystem and the accompanying regulatory and legal framework.

Personalization and Partnership   The essential element of the new application ecosystem that seems to characterize FMC is, fittingly, the new approach to service personalization. As the telecommunication experience becomes more segmented and personal, FMC introduces new tools to support more elaborate application-centric service providers' strategies. For example, the concept of a shared telephone number in the home will be replaced by the concept of public shared identities, where each member of the family will be reachable at his/her own public identity, possibly associated with contact preferences, which include the preferred devices and communication method (e-mail, text message, voice, etc.).

Let's explore the potential benefits of this concept a bit further. While the basic access to the network in the wide area is provided by a common cellular infrastructure (albeit possibly stitched together from networks owned by different service providers, e.g., in the case of roaming), in the home zone a shared broadband connection may be used by a number of customers of different FMC *service* providers competing for business, which may or may not have a partnership in place with the fixed *access* provider.

This is a particularly powerful property of FMC that should not be underestimated. The connection pipe from the home or business premises to the Internet or an operator's IP network belongs to a single fixed access provider, which will inevitably be in a stronger position[6] than its mobile access counterparts or service-only providers to attract customers with its FMC offerings and differentiate its services.

For instance, in a home zone where the fixed-line broadband access connection is run by, for example, AT&T, it will be likely that the most integrated and feature-reach FMC services will be provided by AT&T and its cellular arm or, at the very least, other FMC providers who have been able to negotiate deals to provide FMC services over AT&T's broadband access.

Fixed access providers will be motivated to enter alliances with other service providers and mobile operators as the alternatives (such as creating service offerings themselves or distributing converged handsets on their own with a "bring-your-own mobile access subscription" service model, i.e., without a predetermined mobile service provider subscription) may expose them to various business and technical vulnerabilities. Such vulnerabilities may include less competitive solutions, slower time to market, poor service integration or, in the worst case, motivate their subscribers to switch to competitive fixed access networks that may offer more appealing services and pricing through better conceived partnerships.

Similarly, mobile operators and service providers who do not own access will also come under pressure to partner with fixed access providers. Not doing so will make them vulnerable to competition from fellow mobile carriers as well as service providers from other sectors. For instance, competitors may attempt to lure customers with an FMC service based around the fixed line already available in the premises where their subscribers live or work.

FMCVNO   As we started to outline in the preceding paragraphs, service providers involved in FMC business typically take on two distinct roles.

One role is played by those running fixed or mobile networks offering basic telecommunication services, such as voice or data access, and making their infrastructure available for value-added services provided by others. Despite being in strong position to compete these, operators often find themselves well on the way of becoming utilities, offering an infrastructure generating a predictable and stable level of revenues and income if the subscriber numbers do not fluctuate significantly. Some are perfectly fine with this model. Others, however, are desperately trying to get out.

The other role is played by those dedicated mostly to the business of offering advanced converged *services* over any network, which we identify with the name of "fixed-mobile

---

[6] Potentially even capable of squashing the competition from carriers delivering their services over non-cooperative broadband by applying elaborate QoS and traffic prioritization schemes.

converged virtual network operators" (or FMCVNOs). These operators[7] are already quietly positioning themselves in the growth areas of retail services and content definition, sharing, and distribution. While they will be more exposed to competition and risks, the rewards in venturing into such a new services paradigm will also be greater. This class of service providers will also have greater influence on the evolution of the networks, being closer to the customer and therefore dictating their requirements to the operators in the first role.

It is foreseeable that the large incumbent operators will at first attempt to become FMC service providers by vertically integrating both roles and targeting their large fixed and/or mobile subscriber base. Nevertheless, we believe competitive (and perhaps also regulatory) forces will over time promote the separation of roles; albeit the competitive scenarios in different regions of the world may favor vertically integrated operators over their FMCVNO counterparts. So, FMC's somewhat counter-intuitive side effect is to drive the identification of two new roles in the industry: the role of provider of access services, and the role of provider of services independent of the access service subscribers may choose or happen to be camped on.

Beyond Voice

Like any technology inducing significant change in the current order of things, FMC is expected to spread its influence to other areas of telecom such as near field communication (NFC) messaging, machine-to-machine (M-to-M) communication, and mobile commerce (M-commerce). Next we consider the effect of FMC on those fields in some detail.

FMC + NFC   We will start the discussion of NFC with the introduction of its main underlying technology (or better, its main driver), called radio-frequency identification (RFID). RFID was originally designed to facilitate the exchange of identification and other information between an active RFID party, which normally initiates the ID query transaction and is known as RFID reader, and a passive RFID receiver/transmitter or RFID tag. Early examples of RFID are typically found in badges used to identify personnel, and items (or pallets of items) in stores and warehouses.

RFID technology, when applied to mobile devices, created the foundation for near field communication. Just like RFID, NFC is also a short-range wireless technology based on inductive coupling. NFC operates in the 13.56 MHz frequency range over a typical distance of about four centimeters and supports data rates of up to 424 Kbps. NFC was standardized in International Organization for Standardization (ISO) 18092 [161], ISO 21481 [162], and ETSI TS 102 190 [163] over the course of 2003 and 2004. It is currently governed by the NFC forum, which regularly publishes NFC specifications.

The typical commercial uses for NFC-equipped mobile devices include contact-less credit card–like point of sale (POS) transactions and other proximity payments, certain types of inventory management, and identification for access and entry control.

---

[7] In some instances, different departments or subsidiaries of the same operator offering access services today may take on these two roles.

A scenario of a contact-less payment involves an NFC chip and antenna (NFC tag) embedded in a credit or debit card or an alternative form factor like a keychain token or a mobile phone (some solutions may even allow adding NFC to existing phones by replacing battery covers with an aftermarket accessory including an embedded NFC chip combined with the proper software). The NFC tag in a device or card is dormant until it is placed within a few inches of a compatible POS terminal. The tag and reader then exchange information securely using the radio frequency capability of the system (see Figure 7.1). After that the transactions proceed similarly to today's contact-based card-reading systems.

Obviously security issues are quite relevant when transmission of data over a wireless link is involved, so due consideration as to how to authenticate transactions between reader and tag is definitely in order. One of the challenges is that an RFID tag is by nature a cheap device and the processing power required to run complex ciphering or key exchange algorithms runs counter to the goal of making a device cheap. The area of communication security and mutual authentication between reader and tag to avoid spoofing may be one of the most interesting future areas of investigation and standardization to make NFC useful in many new domains of application.

NFC technology is slowly finding its way into commercial products. The process is slowed down in part due to the classic "chicken and egg" problem stemming from protracted build-out of its ecosystem and industry struggles with business cases and value proposition issues. It is therefore important to consider ways in which NFC can be combined with other technologies such as FMC either to augment the existing solutions by new functionality or even to enable entirely new ones.

NFC can be used as one of the FMC enablers, and in turn can use FMC to expand its value proposition. For one, FMC greatly expands the overall NFC value proposition by increasing back-end infrastructure connection availability thorough the use

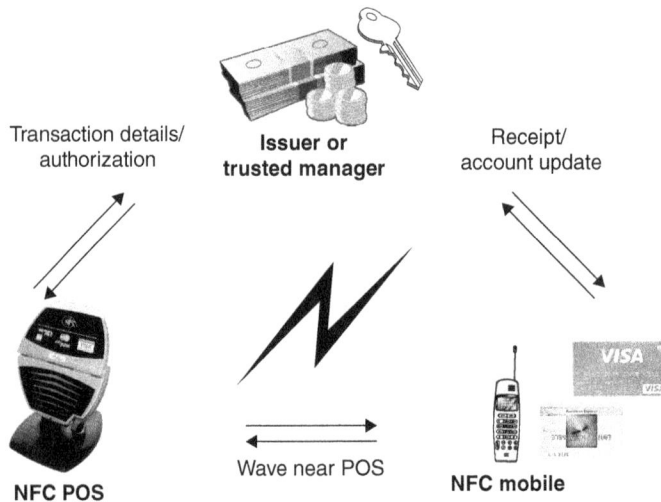

Transaction details/ authorization   **Issuer or trusted manager**   Receipt/ account update

Wave near POS

NFC POS   **NFC mobile**

Figure 7.1   NFC

of multiaccess capability of FMC mobile terminals. In the example with proximity payments, after a POS transaction, the reconciliation with the issuing credit card authority can be conducted via any available fixed or mobile connection, depending on the situation or operator preferences.

Also, NFC-capable terminals may be used to quickly gain access and pay for connectivity at an arbitrary Wi-Fi access point with which there is no pre-established customer/provider relationship instead of conducting a transaction manually by entering credit card information. This is possible, for instance, by placing the FMC/NFC device near a dedicated NFC POS at an airport Wi-Fi hotspot, which will exchange payment information with the hotspot provider, thus transferring (payment) credentials necessary to get access to the service.

In another example, FMC can improve NFC value when used in conjunction with advertising and instant purchases. For example, an NFC/FMC terminal may interact with a billboard advertising the latest musical in town and allow the user to book seats for a show, simply by placing the mobile phone near the billboard, which will cause the terminal to point to the Web site for more information, make seat selections, and then exchange payment information with the site without extensive user manual intervention.

The mobile yet again would access the Internet using the most suitable type of connectivity available in the area (it could be Wi-Fi, WiMax, or cellular). Further, the user would receive a voice mail, an e-mail, an SMS, or any combination of these (depending on user preferences) with confirmation and appropriate instructions, all with greater assurance of multiaccess reachability.

FMC + M-to-M   Improved availability, reachability, and cost savings enabled by FMC can be applied not only to communication between humans or humans and machines, but also to communication between machines themselves, for instance, intelligent devices with microprocessors running applications—machine-to-machine (M-to-M)—with far-reaching consequences.

To date, the solutions offered by telecom service providers have almost always involved a human user at one endpoint of a communication session. With the mass deployment of wireless networks and microprocessor-based remote sensing devices, this paradigm is about to change, with literally millions of devices capable of connectivity being readied by manufacturers in different industries.

Some of the early examples of solutions involving communicating objects include telemetry-based systems monitoring infrastructure integrity and agricultural products and robots in assembly plants, as well as custom-connected home setups similar to the one depicted in Figure 7.2. The solution in Figure 7.2 allows persons and remote applications to monitor the status of stationary and semistationary objects in the home zone and control their behavior based on policies, changing conditions, and other factors. The home-zone objects may interface with the home-zone M-to-M controller via low-power, close-proximity technologies such as ZigBee, defined in IEEE 802.15.4 [164] specifying Wireless Personal Area Network (WPAN), and Z-wave (a proprietary standard developed by a company called Zensys), or more mainstream Bluetooth and Wi-Fi technologies (depending on specific implementations).

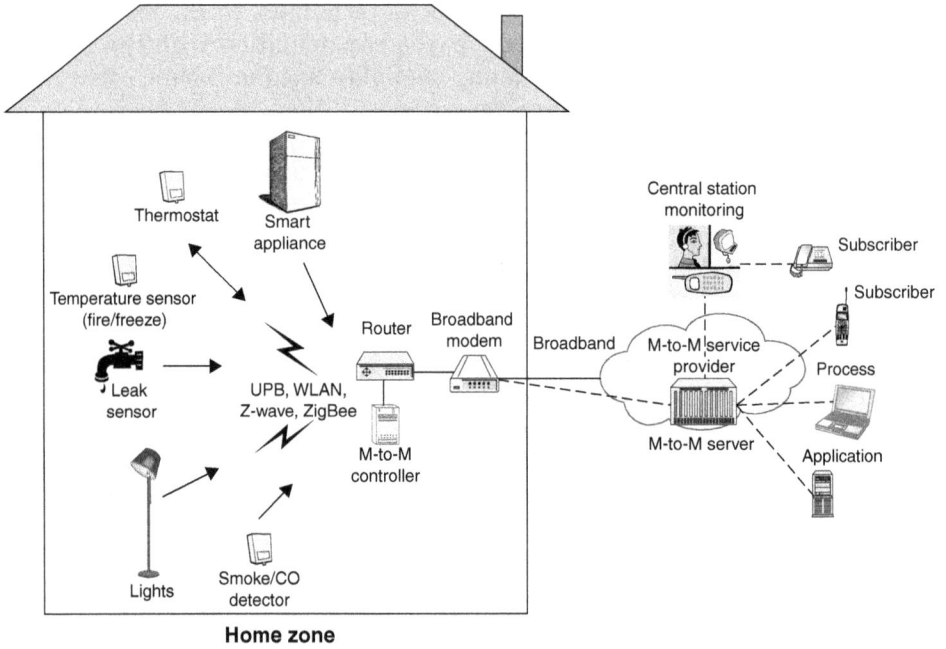

Figure 7.2   Connected home

FMC is a perfect technology to bring mobility to M-to-M solutions. It makes it possible for communicating objects to stay connected while roaming through multiple types of access. The access method selection by such smart machines could be based on availability, cost, time of day, and many other conditions. Such new seamlessly connected types of M-to-M solutions will become more reliable, more cost effective, and easier to deploy in various environments; consequentially, they will offer more value to their users.

The applications relying on such a symbiosis of M-to-M and FMC may include inexpensive and easy-to-install,[8] standards-based, remote monitoring and control equipment. In residential applications, for example, communicating thermostats, security systems, and lighting, as well as numerous mobile assets belonging to a household such as vehicles, pets, and family members, can be enabled to maintain uniform connectivity and create an M-to-M ecosystem, as depicted in Figure 7.3. Other uses for thus defined FMC/M-to-M solutions might include more flexible monitoring of the structural integrity of the public infrastructure and advanced management of alternative energy resources.

---

[8] The appliances and other devices equipped with multimode wireless/wireline interfaces will significantly reduce the installation time and cost.

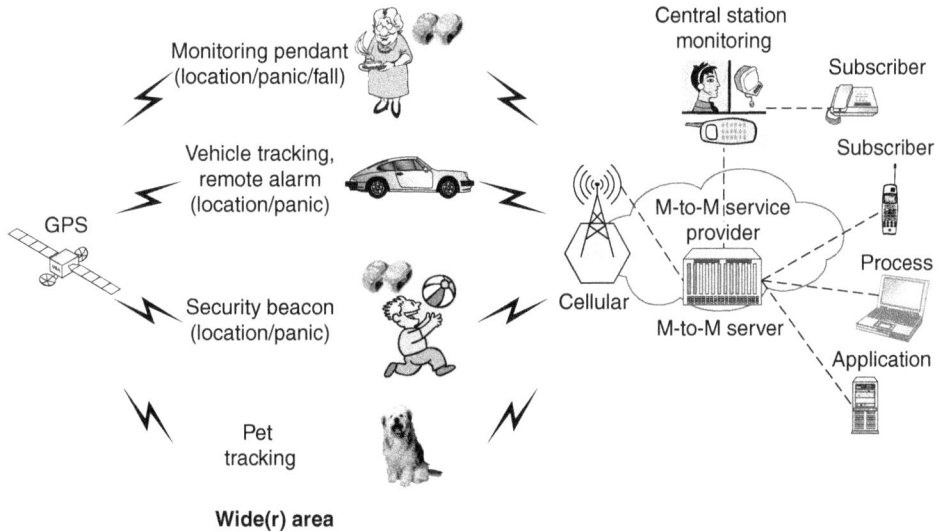

Figure 7.3 M-to-M with FMC: beyond the home zone

FMC + Instant Communications The key difference between instant and traditional voice communications—especially when it comes to the user experience—is, not surprisingly, its immediacy. Mobile instant communication takes many forms, with the most widespread being

- Short Message Service (SMS)
- Mobile instant messaging (IM)
- Push-to-talk (PTT, or as it is increasingly called *push-to-talk over cellular* or PoC)

Immediacy also differentiates most[9] instant messaging solutions from store-and-forward technologies such as e-mail. Indeed, the instant messages sent by one user immediately appear on the other user's device after the session is established, instead of being stored and downloaded from a server by an e-mail client at certain fixed intervals.

Another important feature typically supported in instant communication technologies is presence (described in the previous chapter). For example, IM and PTT users know in advance if the person they desire to communicate with is available. Thus a typical IM user, even when the communication is not fully instant due to network delay or other conditions, is at least offered an *illusion* of real-time conversation, while the e-mail experience remains more akin to sending and receiving physical mail. While SMS most often cannot match PTT and IM in speed, it also provides a near-instant method to reach a person and often comes with immediate delivery confirmation.

---

[9] Some variants of instant messaging also rely on store-and-forward technology.

If only these instant communication services could also become seamless! This is where FMC comes into play. In fact, FMC not only holds the keys to convergence of instant communication systems in fixed and mobile domains, but it also offers the opportunity to create new ways to instantly exchange information regardless of the type of media being used. In fact, as you are going to see further in this section, the process of instant communication convergence has already begun. It could, however, benefit from some harmonization and better standardization to achieve subscriber numbers normally associated with voice services.

PTT over Cellular   Push-to-talk technology brings the experience familiar to the users of walkie-talkie radios to the world of cellular telephony. With walkie-talkies and other similar types of point-to-point radio solutions, the communication session is established instantly upon pushing a designated button and speaking into a speakerphone or a headset. When turned on, the radios usually go in permanent "listening" mode and get "connected"[10] only for as long as required to transmit and receive speech.

Further, the walkie-talkie communication is simplex (only one way at a time), while traditional mobile cellular voice communication is almost always duplex. The push-to-talk over cellular experience is similar, with the PTT handsets behaving comparably to walkie-talkie radios, while the underlying technology is quite different.

The user experience enabled by both walkie-talkie and PoC technology significantly contrasts with modern telephony (both fixed and mobile), where the user must first dial a number (or use a URI with VoIP) and then wait for the dial tone and then for the called party to pick up to start the communication session. The session typically stays on (or packets are streamed end-to-end in case of VoIP) as long as the communicating parties do not hang up.

Since the two types of communications are so fundamentally different, the greatest challenge of PTT commercial deployment in cellular environments was to combine the benefits of both to create a new, unique service. Granted, the resulting PoC experience was not entirely different from that of traditional mobile telephony. After all, there are only so many ways to implement voice communications. However, the PoC unique properties allow cellular service providers to extend their technology to certain communications applications and vertical markets originally not served or supported only by noncellular radio equipment and solutions.

PTT in cellular networks can be implemented in a few different ways, all of which present some sort of a combination of walkie-talkie-like functionality encompassing floor control (who gets to speak at any given moment) with traditional cellular services using data or voice (both control and traffic) channels. The most commercially successful PTT service was introduced by the U.S. company Nextel in the 1990s under the brand of the Direct Connect service running on their Integrated Digital Enhanced Network (iDEN).

---

[10] In a sense, with actual walkie-talkies no real connectivity is ever established. The transmitting radio broadcasts the signal whether or not the receiver is on. If the receiving radio happens to be on, it receives it.

iDEN is a TDM-based, digital-trunked radio system using M16-QAM digital modulation and Vector Sum Excited Linear Predictor (VSELP) speech coding. It is loosely based on GSM and even supports a variant of GSM MAP for signaling, but unlike GSM, being half-duplex it uses only one frequency at a time. It thereby provides significant bandwidth savings in contrast to other cellular systems using separate inbound and outbound frequencies for full-duplex transmission and reception of traffic and signaling. The iDEN user experience is essentially instantaneous, with session establishment (time between pressing the talk button on the sending end and receiving speech on the receiving end) taking less than a second and in-call latency measuring in milliseconds.

While being proprietary and initially not supporting GSM roaming, the iDEN system was custom built to provide PTT service from the ground up and therefore presented a very high barrier of entry for competition. Essentially, a company already engaged in cellular service and wishing to offer PTT—ideally with the ability to roam to and from iDEN—had to recreate a similar system, which proved to be unfeasible from both the business-case and technology perspectives. The industry therefore turned to alternative approaches that would create a solution overlaying the existing networks. In many cases the cellular data networks (at the time, underutilized) lent themselves well to such an overlay, and therefore the technology selected for PTT service was VoIP.

A number of different implementations of VoIP-based, push-to-talk solutions are currently available. One such solution is standardized by 3GPP and OMA (in OMA-AD-PoC [158], OMA-RD-PoC [157]), and other specifications under the name *PTT over Cellular* or PoC. The general architecture of PoC is depicted in Figure 7.4.

OMA PoC views the iDEN user experience as a benchmark. The common measures of performance of a system built to OMA specifications are introduced by OMA under the name Quality of Experience (QoE).

QoE metrics defined in the OMA PoC requirements include

- Right-to-speak (RTS), measuring call setup time or the response times during initial PoC session establishment. RTS measures the time interval between pressing the PoC button and receiving a "right-to-speak" tone.

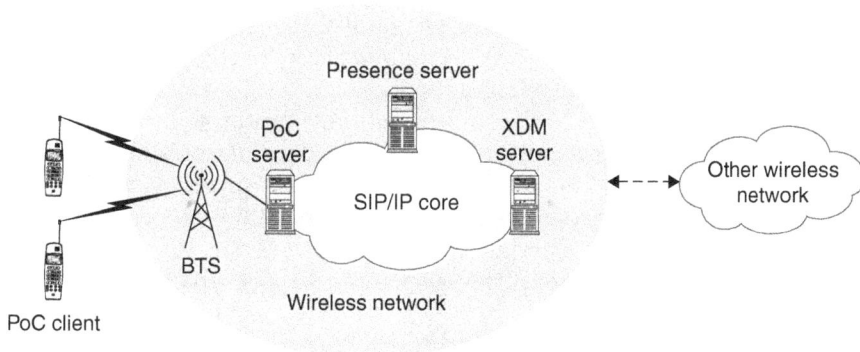

Figure 7.4 PoC architecture

Table 7.1 PTT Performance Measurements

Metric	OMA Guideline	Comparable IDEN Characteristic
QoE 1 Right-to-speak (RTS)	1.6–2 sec	<1.2 sec
QoE2 Start-to-speak (STS)	<1.6 sec	<0.3 sec
QoE3 End-to-end channel delay	First burst: 4 sec subsequent: 1.6 sec	First burst: <1 sec subsequent: <0.3 sec
QoE4 Voice quality	MOS >= 3 at BER <= 2% (MOS being the Mean Opinion Score)	Comparable to standard cellular voice

- Start-to-speak (STS), measuring response time during the established PoC session, or in OMA language "... the duration between the time a PoC participant initiates a floor request (i.e., permission to talk) and when he receives a 'start-to-speak' indication (or queuing indication or denial)."

- End-to-end channel delay, measuring the duration of a time period necessary for speech to propagate from the caller to a called party.

- Voice quality.

Table 7.1 summarizes these OMA QoE parameters and compares them to the equivalent iDEN measures.

At the core of OMA PoC architecture is the OMA PoC server governing both media (flow of RTP packets through the system) and signaling. The examples of signaling include gaining *floor control,* or in OMA terminology "negotiating a permission to send a *Talk Burst,*" [11] managing SIP sessions, and interoperating with other components such as presence servers and the XML Document Management (XDM) function defined in OMA-TS-PoC_XDM-V1 [8].

The OMA PoC specification defined a system virtually independent of the underlying wireless data networks, which therefore can be implemented over many contemporary high-speed cellular networks, such as HSPA or EV-DOrev.A; it can even be extended to Wi-Fi and WiMax in the future. Originally, though, the OMA focus was (and largely still is) on cellular systems, so OMA standards currently do not specify PoC service over noncellular wireless or wireline networks.

Despite this limitation, the underlying PoC protocols lend themselves well to convergence with fixed or any other noncellular data networks and potential interoperation between fixed and mobile PoC endpoints. The appropriate standardization efforts, based on the experience gained from the proprietary experimentation, can round up the picture, defining a foundation and framework for "any-media" PoC service. Thus defined, an all-IP converged PoC service can support roaming over multiple networks

---

[11] Remember that PTT communications are always half duplex. In such an environment, a PoC client can transmit speech only upon gaining floor control.

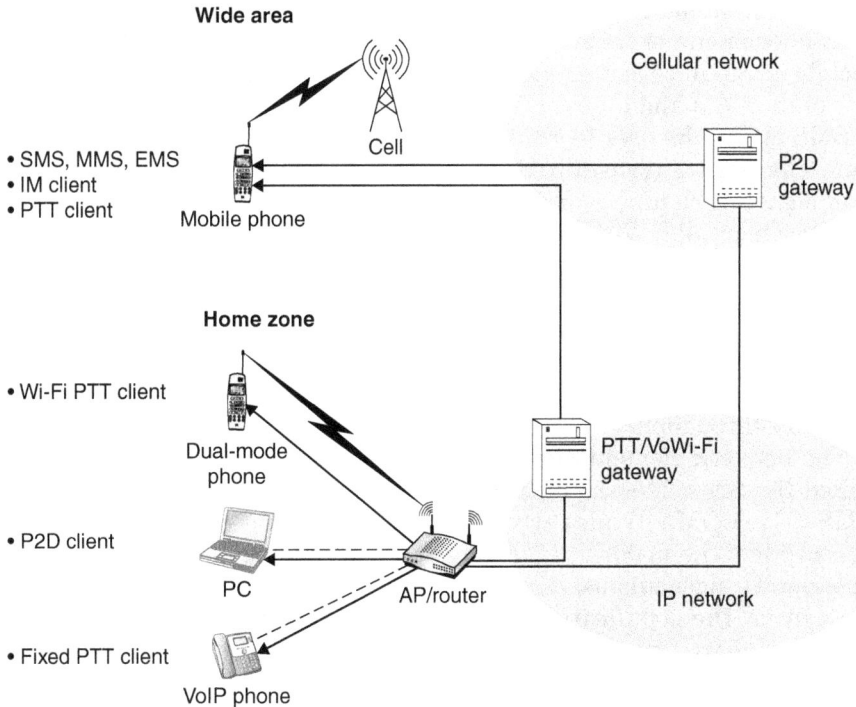

Figure 7.5 Converged PTT solution

and access technologies as long as appropriate data-roaming agreements are in place between different types of carriers.

For example, a desktop PC or fixed VoIP phone, as in Figure 7.5, can be introduced as a potential PoC endpoint, and one can envision gateway solutions allowing interoperation of PoC with other instant communication services such as SMS, IM, or "slower" services like voice mail and even PSTN voice. Moreover, the converged PoC framework can bring intercarrier PoC closer to reality, which will in turn bring it mainstream, driving its penetration rates to extend far beyond the existing vertical markets, just as SMS roaming made SMS what it is today.

Mobile Text Messaging In this section we are going to provide a brief overview of mobile messaging and describe the convergence trends taking place in this technology.

The main forms of mobile messaging include

- Short Message Service (SMS)

- Multimedia Messaging Service (MMS)

- Enhanced Messaging Service (EMS)

- Instant messaging (including popular privately branded services such as AOL, MSN, Yahoo, ICQ, and so on)

The widespread adoption of mobile text messaging has been one of the most significant phenomena of mobile communications in the 1990s. In fact, SMS was the first mobile communication service other than voice to achieve widespread acceptance as one of the GSM and later CDMA offerings.

SMS allows the user to send messages from 140 to 240 bytes, depending on the system. The SMS is transmitted in traditional cellular systems supporting CS services over signaling channels, and therefore can be received and sent while a user is actively engaged in a call.[12] With the advent of GPRS and EVDO, new options became available to transmit SMS over traffic channels.

Unlike IM or OMA PoC, which provides a near-real-time communication experience, the SMS is a store-and-forward technology, which does not guarantee immediate delivery even if a subscriber can be readily located. In practice, however, the user experience is almost always near instant and is only limited by the infrastructure capacity and air interface throughput.

The network element storing SMS messages and managing their flow in GSM is called the Short Messaging Service Center (SMSC) and the SMS Service Center or SMS-SC in CDMA. While early SMS implementations did not provide acknowledgment of reception, the majority of today's solutions support it via a number of standard or proprietary mechanisms.

Early on, the popularity of SMS prompted the industry to look for ways to address its many shortcomings such as limited message size and functionality essentially limited to sending text.[13] Enhanced Messaging Service (EMS) adds to the original SMS standard a number of improvements, such as text formatting options, the ability to send images (albeit limited to 32×32 pixels), animated objects, simple sounds such as ring tones, and finally the ability to merge two or more messages as a way to overcome SMS limits.

While EMS certainly increased SMS usability, a new standard, more suited for today's mobile devices capable of handing multimedia content, was needed. In response to these needs, 3GPP produced a new standard called Multimedia Messaging Service (MMS), which unlike SMS or EMS calls for transmissions over traffic channels, thus inherently supporting the higher bandwidth needed to transmit multimedia content. MMS supports all the features of EMS and SMS, adding the ability to transmit high-resolution color graphics and moving images.

SMS has really taken off globally since carriers started providing SMS roaming support. However, it took the industry almost a decade to recognize and start addressing another major limitation of SMS, namely its mobile-only nature. Unlike voice calls and instant messages, short messages could not be sent between mobile devices and fixed phones or desktop computers.

---

[12] One of the primary functionalities later aimed at by IMS for general data service.

[13] Essentially eliminating both *S*es in the acronym or changing them to something else.

One of the more frequent approaches to addressing the SMS incompatibility with fixed systems involves a gateway converting SMS to e-mail. For example, many carriers nowadays allow any of *their own* mobile subscribers to send messages by using a *number@carrier_name.net* address. Such a message will be directed to a gateway, which then converts it to SMS and forwards it to the specified number. Alternatively, the gateway can redirect the SMS originated from a mobile to a pre-provisioned e-mail address if the destination mobile is not available or its user has not appropriately set his or her preferences. Unfortunately, a common standard for such services does not exist, so users must familiarize themselves with different formats and procedures used by specific carriers.

The problem of SMS messages inadvertently or purposely sent to fixed numbers, which initially caused lost revenue and significant numbers of customer complaints, is now also addressed via a number of proprietary solutions created by the industry for both analog and VoIP fixed environments. For instance, SMS messages sent to analog numbers can be intercepted by a gateway installed in the fixed operator's network and converted to e-mail or made available to subscribers through text-to-speech conversion systems, while a "Message Waiting" notification can be deposited in the VM if the subscriber uses Centrex or some similar service.

When the fixed number is provided through a VoIP service, carriers have more options at their disposal. For example, if the subscriber is using an advanced VoIP phone, then potential feature parity with cellular is possible. That means that the VoIP phone (this also includes VoWi-Fi phones) can be used to send and receive messages similar to a cellular phone down to the user interface. The SMS can be delivered to a VoIP phone using the SIP MESSAGE message method as depicted in Figure 7.6. The same procedure can be applied by a dual-mode phone in Wi-Fi mode. Other options such as forwarding to e-mail are also available. Recently, 3GPP has standardized a solution for the delivery of SMS over any IP-capable means of access, thus making SMS available as a converged service. 3GPP is also in the process of defining the interworking between SMS and other forms of messaging like IM.

Converging Instant Communications   The instant communications convergence is already well underway, with gateways and applications being deployed enabling short messaging and instant messaging between mobile and fixed devices. As various proprietary approaches become interoperable and more integrated, the remaining boundaries between fixed and mobile messaging become more blurred.

These trends in turn may spur new rounds of standardization efforts, establishment of closer ties between different service providers, and eventually, creation of a global messaging network similar to the PSTN of today, which may very well be the final step in creating the elusive *converged or unified messaging* system—a concept of linking all types of messaging under the same service, introduced in the late 1990s.

Certain elements of such a system, like gateways and servers supporting the ability to exchange SMS, MMS, and IM between mobile and fixed devices, are already being

Wide area

• SMS
• EMS
• MMS

Cell

Cellular network

Mobile phone

SMS/MMS/EMS
e-mail gateway

Home zone

• SMS/SIP client
• EMS/SIP client
• MMS/SIP client

Dual-mode
phone

SIP server

SMS/VM
gateway

• SMS/SIP client
• E-mail
• Web-based VM

PC

AP/router

IP network

• SMS/SIP client
• SMS/VM

VoIP phone

• SMS/VM
• MWI
announcement

Analog phone

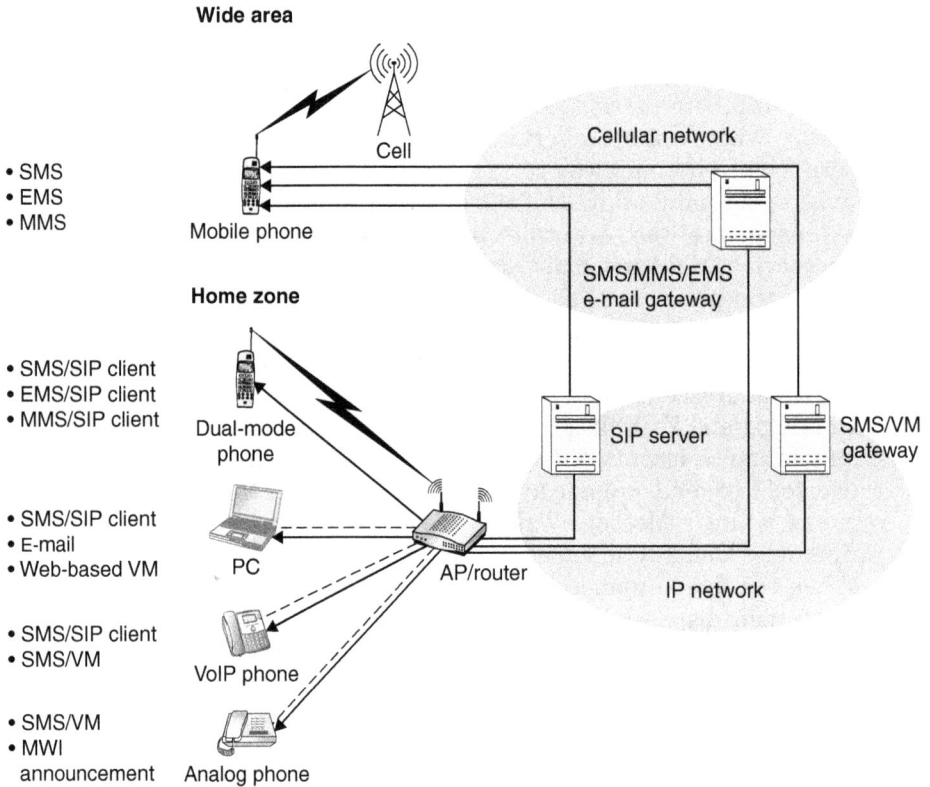

Figure 7.6 Short messaging convergence

put in place by operators creating an early foundation for such systems. The next logical phase would be adding PoC to the mix and enabling the support of all these types
of messaging on the VoIP and dual-mode or multimode FMC terminals via dedicated
bundled clients.

Finally, on the application and user interface levels, service providers could start
introducing the "universal messenger" functionality consisting of the common client
with a standard user interface on the mobile phone, fixed phone, and desktop, or providing Web-based fixed- and mobile-optimized application options. The latter would
extend instant communications to all devices as long as a fast enough data pipe is
available along with a system supporting transcoding of all the message types to formats acceptable by clients on different devices. One possible solution enabling such a
ubiquitous instant communication nirvana in the not-too-distant future is represented
in Figure 7.7.

The use cases of such systems require little explanation. Quite simply, users would
have the full array of instant communication services at their fingertips regardless of
device, location, or type of access, freely selecting the communication method that best
suits their needs at any given moment.

Wide area

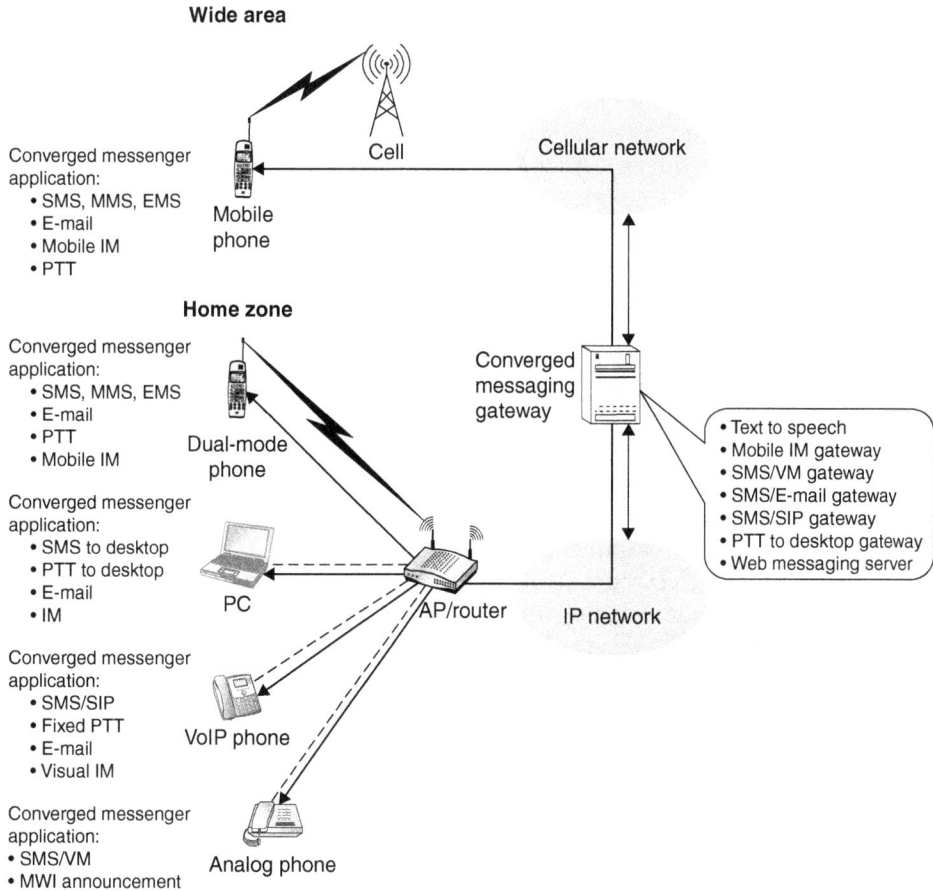

Converged messenger
application:
• SMS, MMS, EMS
• E-mail
• Mobile IM
• PTT

Mobile
phone

Cell

Cellular network

Home zone

Converged messenger
application:
• SMS, MMS, EMS
• E-mail
• PTT
• Mobile IM

Dual-mode
phone

Converged
messaging
gateway

• Text to speech
• Mobile IM gateway
• SMS/VM gateway
• SMS/E-mail gateway
• SMS/SIP gateway
• PTT to desktop gateway
• Web messaging server

Converged messenger
application:
• SMS to desktop
• PTT to desktop
• E-mail
• IM

PC

AP/router

IP network

Converged messenger
application:
• SMS/SIP
• Fixed PTT
• E-mail
• Visual IM

VoIP phone

Converged messenger
application:
• SMS/VM
• MWI announcement

Analog phone

Figure 7.7 Converged instant communications ecosystem

Use Case: A Business Trip

Whether vertically integrated operators or virtual operators prevail in the future of
retail FMC telecommunication services, we can hold great expectations as we look
ahead toward the FMC services applied to both personal and business communications. To provide the gist of what is just around the corner, we will look into a day
in the life of an FMC solution customer.

Alice is a very busy businesswoman. She needs to catch a plane to visit a customer in Paris for a two-day trip, and she will need to catch the Heathrow Express
train early in the morning to get to Heathrow Airport. Due to a sudden failure in

(continues)

the power line, there can be long delays in the Heathrow Express train service. The travel agency has an alert system, which sends an SMS to Alice's mobile, e-mails her, and rings her home phone to alert her in such cases, so, fortunately, she can wake up on time to arrange alternative transportation to the airport.

While at the airport, Alice uses presence and location monitoring application to keep in touch with colleagues bound to join her on the trip. When they arrive at the airport, their mobiles trigger presence status change and proximity alerts, which are promptly delivered to Alice's mobile through the airport Wi-Fi network (Alice's colleagues have authorized Alice to receive information about their location).

Following the reception of these alerts, Alice initiates a conference chat over the Wi-Fi network available at the airport to invite them all to a restaurant in the airport to have a quick breakfast and to sync up just before departing. They also need to conference in some colleagues for the discussion, so they initiate a videoconference on Alice's laptop sharing documents and a whiteboard. The call ends just minutes before boarding, so they need to rush to the gate.

Alice waves her NFC-enabled mobile device above the payment pad located on the table of the restaurant at which they all met, bill is paid and off they go for boarding. The electronic boarding pass was delivered to Alice's mobile, so the NFC-enabled controller at the gate reads the information off Alice's mobile and authorizes boarding. The flight is on time, and they arrive in Paris at the Charles de Gaulle Airport where Alice needs to make a last check of e-mail and have a quick chat with her boss using the local Wi-Fi hotspot.

Both the data and voice connections on her FMC mobile device switch to wide area cellular coverage and continue as she leaves the airport on a taxi, which incidentally was called up by simply waving the mobile near an automatic, taxi-calling, NFC-enabled POS. At the end of the ride, the fare is debited directly to Alice's account via another NFC POS in the cab. During her taxi ride Alice receives an SMS informing her that the foyer light she left on was remotely switched off by the monitoring service.

As they arrive at the customer premises, all the meeting attendees' information is exchanged using an automated NFC-based vCard[14] reader at the reception integrated with the building security system, which verifies the guests' information and transmits guest passes to their mobile devices. At the meeting all the members of Alice's team are on an IM conference session established via a personal area network (PAN), so that they can quickly exchange information as the meeting proceeds.

---

[14] vCard—defined in the Internet Mail Consortium (IMC) vCard specification [160]—is a standard for electronic encoding of personal information such as business cards. A vCard information capsule can be attached to e-mail, distributed over the Internet, or stored in mobile devices and shared through any available wireless network such as Bluetooth PAN.

The meeting went well, and Alice's team goes to dinner. Inside the restaurant (in an old building cellar) there is no wide area cellular coverage, so all communications happen through a femtocell.

While at dinner, Alice and the customers agree it would be a great idea to go for after-dinner drinks at a trendy place in the Latin Quarter. Alice recommends on the fly one of the top-rated jazz clubs using her phone-based navigation system with a rating and recommendation engine. The navigation system in Alice's mobile gives them directions, and they just happen to be a few blocks from there (not by chance, as the application is aware of Alice's location!).

The group has a great time at the club, and Alice organizes taxis. When taxis arrive, Alice receives an alert on her mobile, so they all walk out of the club and customers head for home, Alice and colleagues for the hotel. The address of the hotel is stored on Alice's calendar back on the corporate e-mail server. Alice finds the hotel address and just clicks it to send a location object to the taxi driver's navigation system.

When Alice and colleagues arrive at the hotel, they are all already checked in, as she did it while on the taxi (between writing a meeting report and responding to an IM from her husband). When they enter the hotel, their mobiles are provided with a code that will let them in their rooms by waving their mobiles near the door's handle. This was a long day, and everyone enjoys the night's sleep. All the members of the group are going to be awakened at the same time by their corporate travel calendar application, which reminds them they need to board the plane at 11 A.M. and to pack up on time for that. They all meet up in the lobby for breakfast by agreeing through an IM accessed via the hotel Wi-Fi network (access is granted for free for those that check in remotely before arriving at the hotel and have an NFC-capable terminal), and then they are ready for their flight back.

The interesting thing about this story is that most of the scenarios composing it are based on technology that is already commercially available. They just show examples of how an ecosystem composed of an integrated set of networks and applications can simplify one's life, improve productivity, and even introduce new ways to operate such businesses as restaurants and hotels. Of course the implications for information sharing and privacy, and information access permission control, are all to be investigated and resolved, but certainly the direction is set, as the need is clearly there.

Examples of potential issues with such an ecosystem may be individual privacy and protection of identity, confidentiality, and protection of financial or safety-critical information from spoofing. (What if health insurance companies find out Alice has a stressful life, accompanied by nights out and frequent drinking, if only social, in the process? Or what if some critical information like the access key to enter Alice's room is intercepted and spoofed by thieves?) Solving these issues in a way that keeps the costs of the technology and all the surrounding infrastructure viable represents per se a lucrative area of investigation for commercial implementation and professional services.

While many features of this scenario may not materialize in your city in the course of months, or perhaps even years, due to the inherent inertia of the traditional methods and practices, it is certainly the direction we are heading. We believe that the elements of such a converged ecosystem are already being put into place in some form and are sure to proliferate, once the practical, legal, security, privacy, and regulatory frameworks are in place to make these new converged ways of communication possible.

From Here

In this chapter we have provided a sufficient stimulus to readers willing to engage in further study and development of the new frontier of telecoms that is convergence. FMC and the ecosystem of connectivity, applications, and experience it enables (depicted in Figure 7.8, showing an example of an FMC user-centric, uniformly connected, telecommunications environment) promise to be an exciting area of both business

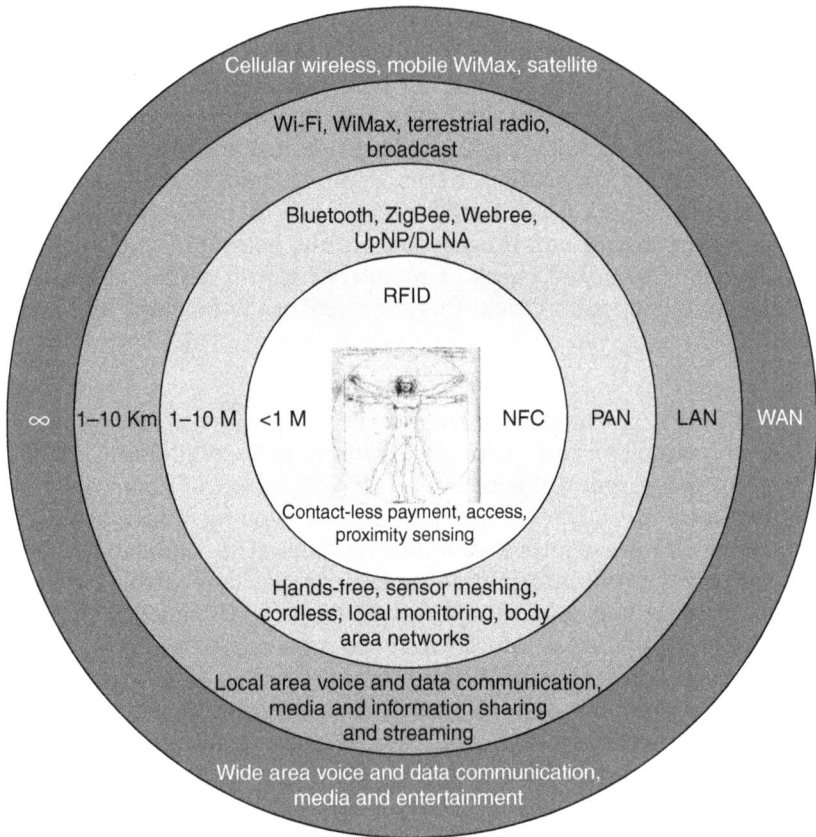

Figure 7.8 FMC ecosystem

and technology, full of new opportunities and challenges. It is also apparent that FMC has the potential to deeply affect (we would like to argue: *improve*) many walks of life and workplaces, by changing the way we communicate and the way humans and machines interact.

It is also equally true that FMC adoption represents a difficult task, as many unexplored regulatory and standards territories need to be entered (e.g., sharing of access networks and content, accessing customer data and subscriber profiles, co-existence of heterogeneous core networks, and monumental security issues stemming from IP connectivity), many technical and business issues need to be resolved,[15] and most importantly, the winning user experience and value propositions must be defined.

Despite these expected growth pains, generally common to any new technology—especially the one of FMC's complexity and magnitude—it is our firm conviction that most of the preconditions for FMC to be successful are there; its technology enablers are being implemented and its value proposition is identified. Most importantly, many of its ecosystem elements are in place, and the demand for FMC services and equipment continues to mount, helping it make the transition from niche technology to a mainstream communication solution.

As we have seen from the example of a day in Alice's life as well as the use-case scenarios and business-case discussions, throughout the book, along with tremendous usefulness for the subscribers, FMC offers hosts of new revenue opportunities for service providers, equipment manufacturers, and other industry players. Moreover, it enables them to enter new markets that until now have remained largely inaccessible and untapped. FMC is also an integral part of the tectonic shifts in the telecom landscape caused by migration of fixed (soon to be followed by mobile) communication to VoIP and IMS accompanied by the invention of new high-speed radio interfaces further helping to bridge the gap between fixed and mobile services.

We hope this book has helped you, the reader, to build at least an initial foundation of the knowledge necessary to understand FMC. That, if anything, should in turn stimulate further investigation and lead your fertile mind to come up with many interesting new applications, services, technologies, and devices, ultimately helping us all get better connected in a more understanding, tolerant, and closely knit society of the future.

---

[15] For instance, the concept of emergency or other mandated services applied differently over each individual access method needs to be replaced by a framework where a more uniform emergency services subsystem is implemented across all types of access. Such a subsystem will need to account for different levels of location information accuracy provided by different networks as well as other distinct properties.

A

FMC Standards

As FMC is about converging fixed and mobile networks, it can be easily imagined that the standards fora traditionally involved in the specification of mobile systems (like 3GPP and 3GPP2, and the WiMAX Forum), those working with wireless access technologies (e.g., the IEEE), and those involved in the specification of fixed networks or applications (like ETSI TISPAN, or the ITU-T Study Groups devoted to Next Generation Networks, and OMA) are quite involved in the industry migration toward converged networks. There are also industry focus groups created to address the interests of industry players involved in convergence, like the Fixed-Mobile Convergence Alliance (FMCA) and the Next Generation Mobile Networks Group (NGMN).

We should also not forget that most of the protocols used in converged networks run over an IP-based transport, and many of these protocols, such as SIP, are standardized by the Internet Engineering Task Force (IETF). The IETF has, therefore, a prominent role in defining today's convergence technologies, especially as the trend in converged networks is toward shifting away from classic SS7 signaling to IETF-defined protocols.

When mobile technologies and applications and handheld devices are involved, the Open Mobile Alliance (OMA) defines standards for instance in the areas of mobile applications and device management. Since FMC is aiming at delivering a seamless, mobile-like, service experience, while blending networks of different kinds, devices, and applications, OMA standards also apply to many FMC aspects beyond applications, and in particular, aspects such as device management (many mobile systems adopt OMA specifications for device management).

While entering into many details about each of these bodies is out of the scope of this book, we feel we need to clarify the general dynamics in FMC standardization activities. We also provide some guidance on how to access standards documentation online for those readers wishing to probe further.

Toward the Identification of Commonalities

The convergence of fixed and mobile systems also implies the convergence or *harmonization* of fixed and mobile standards. In fact, to make the FMC solutions work, standards for fixed and mobile networks need to be carefully assessed for both potential commonalities and incompatibilities. This means that fora that in the past could work quite independently, today are more and more engaged in mutual cooperation and are trying to achieve synergies so as not to duplicate work.

Ideally specifications from different bodies should be kept in sync, to make sure that incompatibilities are identified early enough and thus resolved as part of the normal specification work and new deltas and misalignments are not created during the maintenance and evolution of documents. Such an approach to harmonizing specifications also provides benefits of scale that can be derived from using common architectures and standards for the core networks of converged fixed and mobile systems.

In general, the need for standards synchronization is driving the industry as a whole toward the consolidation of standard bodies and the definition of common interfaces and architectures applicable for both mobile and fixed networks. The best example of one such architecture, known in standards jargon as *common IMS* or *common core,* is being actively worked on by 3GPP, as this organization has been at the forefront of the IMS standardization. This process is now expanding to involve not only all the organizational partners of 3GPP—which as part of their mandate also define regional standards for fixed-line networks—but also the 3GPP2 and CableLabs representing interests of cable operators.

Standards Organizations' Background

In order to understand these dynamics better, it is useful for us to provide some background on the standards bodies and organizations involved in this process.

3GPP and 3GPP2

The Third-Generation Partnership Project (3GPP) is a collaboration agreement that was established in December 1998 among

- ETSI (European Telecommunications Standards Institute), representing Europe
- ARIB (Association of Radio Industries and Business) and TTC (Telecommunication Technology Committee), representing Japan
- CCSA (China Communications Standards Association), representing China
- TTA (Telecommunications Technology Association), representing South Korea
- ATIS (Alliance for Telecommunications Industry Solutions), for North America

This collaboration was aimed at the definition of a global *third-generation* cellular standard, based on the evolution of the GSM core network. 3GPP standards are organized as *releases*. The IMS was defined in 3GPP Release 5, and since then there have been more releases of 3GPP standards and therefore more releases of IMS standards, each characterized by an additional feature set. 3GPP not only defines standards for the 3G system and its evolution but also has taken on the maintenance role of GSM specifications from ETSI and works on the evolution of the GSM system and the interworking of GSM with the 3G system.

With the same goal as 3GPP, but based on the evolution of the IS-41-based CDMA core network, the 3GPP2 was created in 1998 as a collaboration agreement among

- ARIB/TTC (Japan)
- CCSA (China)
- TIA (North America)
- TTA (South Korea)

Following the 3GPP lead, the 3GPP2 has adopted the IMS under the name MMD. 3GPP2 has always worked closely with 3GPP to fine-tune IMS to the needs of CDMA markets. Interestingly, it is becoming clear that 3GPP and 3GPP2 are also on a convergence path with regard to the mobile system specifications and evolution. In fact, some 3GPP2 operators are increasingly considering technologies also selected by 3GPP, such as the evolved 3GPP UTRA, in addition to traditional CDMA radio interfaces. The mechanisms to permit a dual-mode 3GPP2/Evolved 3GPP UTRA handset handover between the 3GPP2 legacy and the evolved 3GPP system are also being defined.

This trend will eventually result in the emergence of the commercial availability of a single global system for cellular mobile communications as early as in the next decade.

ETSI

ETSI TISPAN (standing for Telecoms and Internet Converged Services and Protocols for Advanced Networks) is an ETSI group specializing in fixed networks and convergence. It is no surprise then that it has become such an important forum and a reference for the migration of fixed networks to IP-based transport and, in particular, toward IMS-based architectures (also known as Next-Generation Network, or NGN).

TISPAN's focus is to define the European view of NGN. Similar to 3GPP, TISPAN work is also organized in releases. TISPAN NGN-Release 1 was published in December 2005. TISPAN is currently in a phase where some of its work items related to IMS are being transferred to 3GPP, as part of the concentration of common IMS work in 3GPP.

ITU-T

The ITU-T is the Telecommunication Standardization Sector of the International Telecommunications Union (ITU), an agency of the United Nations. The ITU—the oldest of all telecommunications standards bodies—was founded as the International Telegraph Union in Paris in May 17, 1865. Its roles today include standardization, allocation of radio spectrums, and organizing international telephone interconnection.

The ITU-T was also known as CCITT, from the French name "Comité consultatif international téléphonique et télégraphique." The ITU-T has been host to initiatives to define the future IP-based NGN since 2003, and Recommendations Y.2001 [188] and Y.2011 [189] are the basis for NGN studies and work in ITU-T. In June 2004, this work resulted in the creation of a Focus Group on NGN (FGNGN) with the aim of globalizing the NGN standards, which so far had been developed fairly independently by the regional Standards Development Organizations (SDOs).

The results of this work were transferred to the ITU-T Study Group (SG) 13, which is the lead study group for NGN in ITU-T, and SG 11. Currently the ITU-T undertakes NGN work as the NGN Global Standards Initiative (NGN-GSI).

IEEE and the WiMAX Forum

The Institute of Electrical and Electronic Engineers (IEEE) is developing standards for wireless access technologies such as WiMAX (802.16) and Wi-Fi (802.11).

The IEEE is an association of engineers that among other initiatives—including organizing conferences and promoting professional development of its members—also develops standards, the most well known being the family of 802.x standards related to local area networking, including

- IEEE 802.3 (also more commonly known as Ethernet)
- IEEE 802.11x (known as Wi-Fi or Wireless LAN)
- IEEE 802.16 (known as WiMAX)

Many wireline and wireless operators today are expanding the technologies they use to serve their customers to include Wi-Fi and/or WiMAX service. WiMAX in particular has its own forum (called in fact the WiMAX Forum) to specify the system aspects of deploying WiMAX in a carrier network. To this effect, the WiMAX Forum works closely with service providers (many of which are also members of the forum) and regulators to ensure that WiMAX Forum–certified systems meet customer and government requirements.

The WiMAX Forum also certifies the compatibility and interoperability of broadband wireless products based on the IEEE 802.16 standard. The certification is based on the WiMAX forum system specification. The WiMAX forum has released its specifications Release 1.0 and is working on other releases (starting from the release known as Release 1.5). The WiMAX Forum also adopts the IMS as its multimedia core subsystem and cooperates with 3GPP to make IMS well integrated in the WiMAX system and with both 3GPP and 3GPP2 to enable interworking and handovers.

IETF

The Internet Engineering Task Force (IETF) is a large open international community of engineers contributing their experience gained as network designers, network operators, system or equipment designers, and researchers toward the smooth evolution of the Internet architecture and the smooth operation of the Internet. It is open to any interested individual. There is no concept of IETF membership or company membership, so anyone can come to any of the IETF meetings and anyone is able to participate in the IETF standardization work, which primarily happens via discussion on mailing lists. In practice however, many of the IETF meeting attendees and mailing list discussion contributors belong to companies and other organizations that sponsor their attendance.

The technical work of the IETF is done by working groups, which belong to an area (e.g., the Internet Area, the Routing Area for topics related to exchange of routing information among routers and hosts, the Transport Area for topics related to transport protocols and QoS, etc.). As already mentioned, the IETF work happens primarily via a protocol defined by the IETF itself (that is, via e-mail sent to discussion lists).

The result of the IETF work on a specific topic is a "Request for Comments" (RFC), which is a text document either defining a protocol, an architecture, and a best practice or defining the applicability of a protocol (e.g., an applicability statement about a routing protocol) or an architecture. An RFC can be classified as standards track or just informational. There are also experimental documents that over time may turn into standards if justified by the level of adoption in the Internet.

Typically a standards track document undergoes an evolution from "proposed standard" to "draft standard" to "standard," depending on the number of interoperable implementations, their stability, and their adoption in the Internet. In particular, no protocol can become a standard if there is no evidence of implementations that interoperate based on adherence to the text in this particular RFC. Running code is an essential part of the IETF specification effort; consensus does not have to be unanimous but is "sensed" by the WG chair, who can apply a lot of judgment in the process of defining IETF protocols.

Some of the most important examples of the IETF work, besides the definition of the Internet Protocol itself and the TCP/IP and UDP/IP transport protocol suites, have been the definition of the SIP protocol (adopted for IMS sessions control), and the DIAMETER protocol for AAA. DIAMETER is used for interaction with servers within the IMS, e.g., to allow interfacing with the HSS (Home Subscriber Server) and the PCC (Policy Control and Charging) infrastructure.

The IETF is working closely with most of the standards fora to make sure the evolution of the Internet and IP-based protocols meets the ever-changing requirements of the industry, and it is for this reason that 3GPP, ITU-T, 3GPP2, and other fora have established liaisons with the IETF.

OMA

The OMA (Open Mobile Alliance) has for its mission "to facilitate global user adoption of mobile data services by specifying market-driven mobile service enablers that ensure service interoperability across devices, geographies, service providers, operators, and networks, while allowing businesses to compete through innovation and differentiation."

This means that the OMA defines mobile applications that do not depend on or make any assumptions about the underlying access networks being used, thus being perfectly aligned with the need of delivering converged applications in converged networks. The OMA also conducts interoperability events to verify that specifications OMA generates are standards of high quality. So the completion status of a specification is also associated with interoperability events, not unlike the IETF. The examples of standards developed by OMA include mobile device management (also know as OMA-DM), mobile presence, and push-to-talk over cellular (PoC) service.

FMCA

The recently established Fixed-Mobile Convergence Alliance (FMCA) is a global alliance of telecom operators whose objective is to accelerate the development of convergence products and services. The FMCA mission statement is quite telling: "To abolish the distinction between fixed and mobile for our customers, providing superior wireless services irrespective of the underlying fixed or mobile networks."

To accomplish its goals, the FMCA seeks collaboration with SDOs and the vendor community.

NGMN

The Next-Generation Mobile Networks (NGMN) initiative, which is led by the CTOs of major mobile and converged operators on the planet, intends to act as a catalyst for the ongoing work in standardization bodies on the evolution of mobile systems "by providing a coherent view of what the operator community is going to require in the decade beyond 2010."

The vision of NGMN is one integrated network for the seamless introduction of mobile broadband services, while coexisting with other networks or migrating from other networks. The ultimate goal of NGMN operators is the commercial launch of a new solution in mobile broadband communications that will drive the creation of a new mobile broadband ecosystem.

The target architecture that NGMN operators and affiliated vendors will attempt to propagate in various fora is an optimized packet-switched (PS) network architecture, which offers improved cost competitiveness with alternate types of access and is characterized by higher data rates as compared to existing 3G networks (UMTS HSPA and CDMA DOrA). This architecture will enable a personalized broadband access experience and consolidate in a single packet core the different technologies operated by mobile operators.

Concluding Remarks

In this section we have provided only a brief overview of some of the players involved in the definition for FMC standards, which nevertheless makes it clear that this is a process involving numerous organizations and spanning whole systems from handsets and radio interfaces to core networks and applications. These various players are currently working toward reaching ways to harmonize their specifications by identifying areas of common standardization activity to be concentrated in a single forum. This is expected to drive consistency in definitions of interfaces and protocols and to lead to the elimination of redundancies in the efforts implied by maintaining the specifications of converged systems operations in different fora.

Finding Standards Documents Online

All of the bodies described here make their work available online on their Web sites according to their own policies in terms of free or fee/license-based access. In Table A.1 we provide pointers to the Web sites for readers interested in probing further.

TABLE A.1   Web Site Links for Some Standards Fora Involved in Aspects of FMC

Forum	Web Site
3GPP	www.3gpp.org/
3GPP2	www.3gpp2.org/
TISPAN	www.etsi.org/tispan/
ITU-T	www.itu.int/ITU-T/
IUT-T NGN GSI	www.itu.int/ITU-T/ngn/index.phtml
IETF	www.ietf.org/
IEEE	www.ieee.org/
WiMAX forum	www.wimaxforum.org/
OMA	www.openmobilealliance.org/
The FMCA	www.thefmca.com/
NGMN	www.ngmn.org/

Acronyms

3GPP	Third-Generation Partnership Project
3GPP2	3GPP Two
AAA	Authentication, Authorization, and Accounting
AES	Advanced Encryption Standard
ALG	Application Level Gateway
AMPS	Advanced Mobile Phone Service
AMR	adaptive multirate
AP	access point
APN	access point name
ARIB	Association of Radio Industries and Businesses
ARPU	average revenue per user
ASN	Access Service Network
ASNGW	ASN gateway
ATIS	Alliance for Telecommunications Industry Solutions
ATM	Asynchronous Transfer Mode
B2BUA	back-to-back user agent
BGCF	breakout gateway control function
BICC	Bearer Independent Call Control
BICSN	Bearer-Independent Circuit-Switched Network
B-ISDN	Broadband ISDN
BRA	basic rate access
BS	base station
BSC	base station controller
BSS	base station subsystem
BTS	base transceiver station
CAMEL	Customized Applications for Mobile Network Enhanced Logic

CAP	CAMEL Application Part
CAPEX	capital expenditure
CBR	constant bit rate
CCA	Controlled Channel Access
CCITT	Comité Consultatif International Téléphonique et Télégraphique
CCSA	China Communications Standards Association
CDMA	Code Division Multiple Access
CF	conditional forwarder
CPE	customer-premises equipment
CS	circuit-switched
CSCF	Call Session Control Function
CSMA-CA	Carrier Sense Multiple Access–Collision *Avoidance*
CSMA-CD	Carrier Sense Multiple Access–Collision *Detection*
CSN	Connectivity Service Network
CT2	Cordless Telephony 2
CTP	Cordless Telephony Profile
DECT	Digital Enhanced Cordless Telecommunications
DHCP	Dynamic Host Configuration Protocol
DLE	Digital Local Exchange
DNS	Domain Name System
DR	Dual Radio
DSCP	Differentiated Services Code Points
DSL	Digital Subscriber Line
DSSS	direct-sequence spread spectrum
DTF	Domain Transfer Function
DTM	Dual Transfer Mode
DTMF	dual-tone multifrequency
EAP	Extensible Authentication Protocol
EAP-SIM	EAP-Subscriber Identity Module
EAP-TLS	EAP-Transport Layer Security
ECMA	European Computer Manufacturers Association
EDCA	Enhanced Distributed Channel Access
EDGE	Enhanced Data Rates for GSM Evolution
EF	Expedite Forwarding
EMS	Enhanced Message Service
ENUM	E.164 Number and DNS
E-OTD	Enhanced Observed Time Difference
ESP	Encapsulating Security Payload
ETACS	Extended TACS
ETSI	European Telecommunication Standards Institute

EV-DO	Evolution – Data Only (or Data Optimized)
FA	Foreign Agent
FDD	frequency-division duplexing
FHSS	frequency-hopping spread spectrum
FMC	fixed-mobile convergence
FMCA	Fixed-Mobile Convergence Alliance
FMCVNO	fixed-mobile converged virtual network operators
FQDN	Fully Qualified Domain Name
FTTC	fiber to the curb
FTTP	fiber to the premises
GA	generic access
GA-CSR	GA Circuit Switched Resource
GAN	Generic Access Network
GANC	GAN controller
GA-PSR	GA packet-switched resource
GA-RC	GA Resource Control
GERAN	GSM/EDGE Radio Access Network
GGSN	gateway GPRS support node
GMLC	gateway MLC
GMSC	gateway MSC
GMSK	Gaussian minimum-shift keying
GPRS	General Packet Radio Service
GPS	Global Positioning System
GRX	GPRS Roaming Exchange
GSM	Global System for Mobile Communications
GTP	GPRS Tunneling Protocol
HA	Home Agent
HCCA	Hybrid-Coordinated Controlled Channel Access
HCF	Hybrid Coordination Function
HLR	Home Location Register
HRPD	High-Rate Packet Data
HSDPA	High-Speed Downlink Packet Access
HSPA	High-Speed Packet Access
HSS	Home Subscriber Server
HSUPA	High-Speed Uplink Packet Access
HTTP	Hypertext Transfer Protocol
IBCF	Interconnection Border Control Function
IBM	International Business Machines
ICE	Interactive Connectivity Establishment
I-CSCF	Interrogating CSCF

ID	industrial design
iDEN	Integrated Digital Enhanced Network
IETF	Internet Engineering Task Force
ILEC	incumbent local exchange carrier
IM	instant messaging
IMPP	Instant Messaging and Presence Protocol
IMRN	IMS routing number
IMS	IP Multimedia Subsystem
IMSI	International Mobile Subscriber Identity
IN	Intelligent Network
INAP	Intelligent Network Application Part
IP	Internet Protocol
IPsec	IP Security
IPTV	Internet Protocol–based TV
IS	Interim Standard
ISDN	Integrated Services Digital Network
ISIM	IMS SIM
ISM	Industrial, Scientific, and Medical
ISO	International Organization for Standardization
ISO-OSI	ISO Open Systems Interconnection
ISUP	ISDN User Part
ITU	International Telecommunications Union
ITU-T	International Telecommunications Union–Telecommunication Standardization Sector
IVR	interactive voice response system
LAN	local area network
LBS	location-based services
LCS	Location Services
LSP	label-switched path
LTE	Long Term Evolution
M3UA	MTP3 User Adaptation Layer
MAP	Mobile Application Part
MBONE	Multicast Backbone
MEGACO	Media Gateway Control Protocol (also known as H.248)
MGCF	MGW Control Function
MGW	media gateway
MH	multiple handsets
MIME	Multipurpose Internet Mail Extension
MIMO	multiple-input multiple-output
MIN	Mobile Identity Number

MIP	Mobile IP
MLC	Mobile Location Center
MM	mobility management
MMD	Multimedia Domain
MME	mobility management entity
MMS	Multimedia Messaging Service
MN	multiple numbers
MPLS	Multi-Protocol Label Switching
MRF	Media Resources Function
MRFC	Multimedia Resources Function Controller
MRFP	Multimedia Resources Function Processor
MS	mobile station
MSC	Mobile Switching Center
MSID	Mobile Subscriber Identity
MSISDN	MS international ISDN number
MSRN	Mobile Station Roaming Number
MTP-1	Message Transfer Part layer 1
MTP-2	Message Transfer Part layer 2
MTP-3	Message Transfer Part layer 3
MVNO	mobile virtual network operator
NAPTR	Naming Authority Pointer
NAT	Network Address Translation
NFC	near field communication
NGMN	Next-Generation Mobile Networks
NGN	Next-Generation Network
NIC	network interface card
NICC	Network Interoperability Consultative Committee
NMT	Nordic Mobile Telephone
OAM&P	operation administration, management, and provisioning
OCx	optical carrier level x
OFDMA	orthogonal frequency division multiple access
OMA	Open Mobile Alliance
OPEX	operational expenditure
OSA	Open Service Architecture
PABX	private automatic branch exchange
PBX	private branch exchange
PC	personal computer
PCC	Policy Control and Charging
PCEF	Policy and Charging Enforcement Function
PCF	packet control function

PCM	Pulse Coded Modulation
PCRF	Policy and Charging Rules Function
P-CSCF	Proxy CSCF
PDA	personal digital assistant
PDC	Personal Digital Cellular
PDH	Plesiochronous Digital Hierarchy
PDP	Packet Data Protocol
PDSN	packet data serving node
PGW	PDN gateway
PHB	per-hop behavior
PHS	Personal Handy-phone System
PMIP	Proxy MIP
PoC	PTT over Cellular
POS	point of sale
POTS	plain old telephone service
PRA	primary rate access
PS	packet-switched
PSI	Public Service Identity
PSTN	Public Switched Telephone Network
PTT	push-to-talk
QoS	Quality of Service
RACF	Resource Admission Control Function
RACS	Resource Admission Control Subsystem
RAN	radio access network
RF	radio frequency
RFC	Request for Comments
RFID	radio-frequency identification
RNC	radio network controller
RRA	Radio Resource Agent
RRC	Radio Resource Control
RRM	Radio Resources Management
RSVP	Resource Reservation Protocol
RTCP	RTP Control Protocol
RTP	Real-Time Transport Protocol
R-UIM	Removable User Identity Module
SAE	System Architecture Evolution
SAP	Session Announcement Protocol
SBC	session border controller
SCCP	Signaling Connection Control Part
SCIM	Service Capability Interaction Manager

SCP	service control point
S-CSCF	Serving CSCF
SCTP	Stream Control Transmission Protocol
SDH	Synchronous Digital Hierarchy
SDP	Session Description Protocol
SEGW	Security Gateway
SGSN	Serving GPRS Support Node
SGW	serving gateway
SH	single handset
SIGTRAN	Signaling Transport
SIM	Subscriber Identity Module
SIP	Session Initiation Protocol
SLA	Service Level Agreement
SLF	Subscriber Locator Function
SMS	Short Message Service
SN	single number
SNA	Systems Network Architecture
SONET	Synchronous Optical Networking
SR	Single Radio
SRTP	Secure RTP
SS No. 7	Signaling System Number 7
SSID	service set identifier
SSP	service switching point
STM-x	Synchronous Transport Module level x
STP	signaling transfer point
STS-x	Synchronous Transport Signal level x
STUN	Simple Traversal of User Datagram Protocol (UDP) Through NATs
TACS	Total Access Communication System
TAM	total addressable market
TCAP	Transaction Capabilities Application Part
TCP	Transmission Control Protocol
TDD	time-division duplexing
TDM	time-division multiplexing
TDMA	time division multiple access
TD-SCDMA	time division–synchronous code division multiple access
TFO	tandem-free operation
TIA	Telecommunications Industry Association
TISPAN	Telecoms & Internet Converged Services & Protocols for Advanced Networks
TRAU	transcoding unit

TrFO	Transcoder-Free Operation
TrGW	transition gateway
TSG	Technical Study Group
TTA	Telecommunications Technology Association
TTC	Telecommunications Technology Committee
TUP	Telephone User Part
TURN	Traversal Using Relay NAT
TV	television
UA	user agent
UAC	UA client
UAS	UA server
UDP	User Datagram Protocol
UE	user equipment
UGS	Unsolicited Grant Service
UMA	Unlicensed Mobile Access
UMB	Ultra Mobile Broadband
UMTS	Universal Mobile Telecommunications System
URI	Uniform Resource Identifier
USIM	Universal SIM
U-TDOA	Uplink Time Difference of Arrival
UTRAN	UMTS Terrestrial Radio Access Network
VBR	variable bit rate
VCC	Voice Call Continuity
VDI	VCC domain transfer URI
VLR	Visitor Location Register
VM	voice mail
VoIP	Voice over IP
WAN	wide area network
WB-AMR	Wideband AMR
W-CDMA	Wideband CDMA
WEP	Wired Equivalent Privacy
Wi-Fi	Wireless Fidelity
WIN	Wireless IN
WLAN	wireless LAN
WPA	Wi-Fi Protected Access
WWW	World Wide Web
XML	Extensible Markup Language
XMPP	Extensible Messaging and Presence Protocol

[1] C. E. Shannon, "A Mathematical Theory of Communication," *Bell System Technical Journal,* vol. 27, pp. 379–423, 623–656, July, October 1948

[2] ITU-T Recommendation I.430, "ISDN Basic User-Network Interface – Layer 1 specification," November 1995

[3] ITU-T Recommendation I.431, "Primary Rate User-Network Interface – Layer 1 specification," June 1996

[4] ITU-T Recommendation I.420, "Basic User-Network Interface," November 1988

[5] ITU-T Recommendation I.421, "Primary Rate User-Network Interface," November 1988

[6] ITU-T Recommendation G.711, "Pulse Code Modulation (PCM) of Voice Frequencies," November 1988

[7] *Engineering and Operations in the Bell System, Second Edition,* AT&T Bell Laboratories, 1983

[8] ITU-T Recommendation G.707, "Network Node Interface for the Synchronous Digital Hierarchy (SDH)," October 2000

[9] ANSI T1.105, "Telecommunications – Synchronous Optical Network (SONET) – Basic Description Including Multiplex Structures, Rates, and Formats," 1995

[10] ITU-T Recommendation E.164, "The International Public Telecommunication Numbering Plan," February 2005

[11] ITU-T Recommendation E.123, "Notation for National and International Telephone Numbers, E-Mail Addresses, and Web Addresses," February 2001

[12] ITU-T Recommendation Q.1901, "Bearer Independent Call Control Protocol, CS1," June 2000

[13] ITU-T Recommendation Q.1902, "Bearer Independent Call Control Protocol, CS2," January 2006

[14] ISUP Q.761 to Q.764 T1.113 EN 300 356-1 SPEC/007 N/A N/A

[15] ITU-T Recommendation Q.1208, "General Aspects of the Intelligent Network Application Protocol," September 1997

[16] 3GPP TS 29.002, "Mobile Application Part (MAP) Specification"

[17] ITU-T Recommendation Q.767, "Application of the ISDN User Part of CCITT Signalling System No. 7 for International ISDN Interconnections," February 1991

[18] ITU-T Recommendation Q.761, "Signalling System No.7 – ISDN User Part Functional Description," September 1997

[19] ITU-T Recommendation Q.762, "Signalling Systems No.7 – ISDN User Part General Functions of Messages and Signals," September 1997

[20] ITU-T Recommendation Q.763, "Signalling System No. 7 – ISDN User Part Formats and Codes," September 1997

[21] ITU-T Recommendation Q.764, "Signalling System No. 7 – ISDN User Part Signalling Procedures," September 1997

[22] ITU-T Recommendation G.722, "7 kHz Audio-Coding Within 64 kbit/s," November 1988

[23] IETF RFC3261, "SIP: Session Initiation Protocol," J. Rosenberg et al., June 2002

[24] IETF RFC3525, "Gateway Control Protocol Version 1," C. Groves et al., June 2003

[25] ITU-T Recommendation H.248-1, "Gateway Control Protocol: Version 3," September 2005

[26] IETF RFC3588, "Diameter Base Protocol," P. Calhoun et al., September 2003

[27] IETF RFC3550, "RTP: A Transport Protocol for Real-Time Applications," H. Schulzrinne et al., July 2003

[28] IETF RFC4566, "SDP: Session Description Protocol," M. Handley et al., July 2006

[29] IETF RFC2974, "SAP Session Announcement Protocol," M. Handley et al., October 2000

[30] IETF RFC3966, "The tel URI for Telephone Numbers," H. Schulzrinne, December 2004

[31] IETF RFC2045, "Multipart Internet Mail Extensions (MIME) Part One: Format of Internet Message Bodies," N. Freed et al., November 1996

[32] IETF RFC3204, "MIME Media Types for ISUP and QSIG Objects," E. Zimmerer et al., December 2001

[33] ITU-T Recommendation H.323, "Packet-Based Multimedia Communications Systems," June 2006

[34] IETF RFC2976, "The SIP INFO Method," S. Donovan, October 2000

[35] IETF RFC3262, "Reliability of Provisional Responses in the Session Initiation Protocol (SIP)," J. Rosenberg et al., June 2002

[36] IETF RFC3311, "The Session Initiation Protocol (SIP) UPDATE Method," J. Rosenberg et al., June 2002

[37] IETF RFC3265, "Session Initiation Protocol (SIP)–Specific Event Notification," A. B. Roach, June 2002

[38] IETF RFC3903, "Session Initiation Protocol (SIP) Extension for Event State Publication," A. Niemi et al., October 2004

[39] IETF RFC2475, "An Architecture for Differentiated Services," S. Blake et al., December 1998

[40] IETF RFC2205, "RSVP Resource ReSerVation Protocol (RSVP) – Version 1 Functional Specification," R. Braden et al., September 1997

[41] ITU-T Recommendation Q.1912.5, "Interworking Between Session Initiation Protocol (SIP) and Bearer Independent Call Control Protocol or ISDN User Part," March 2004

[42] 3GPP TS 29.163, "Interworking Between the IP Multimedia (IM) Core Network (CN) Subsystem and Circuit Switched (CS) Networks"

[43] 3GPP TS 29.162, "Interworking Between the IM CN Subsystem and IP Networks"

[44] IETF RFC2960, "Stream Control Transmission Protocol," R. Stewart et al., October 2000

[45] IETF RFC3332, "Signaling System 7 (SS7) Message Transfer Part 3 (MTP3) – User Adaptation Layer (M3UA)," G. Sidebottom et al., September 2002

[46] IETF RFC3761, "The E.164 to Uniform Resource Identifiers (URI) Dynamic Delegation Discovery System (DDDS) Application (ENUM)," P. Faltstrom et al., April 2004

[47] IETF RFC3489, "STUN – Simple Traversal of User Datagram Protocol (UDP) Through Network Address Translators (NATs)," J. Rosenberg et al., March 2003

[48] "Interactive Connectivity Establishment (ICE): A Methodology for Network Address Translator (NAT) Traversal for Offer/Answer Protocols," J. Rosenberg et al., Work in Progress, October 2006

[49] "Obtaining Relay Addresses from Simple Traversal Underneath NAT (STUN)," J. Rosenberg, Work in Progress, October 2006

[50] IETF RFC2474, "Definition of the Differentiated Services Field (DS Field) in the IPv4 and IPv6 Headers," K. Nichols et al., December 1998

[51] IETF RFC3270 "MPLS Support of Diff-Serv," F. Le Faucheur et al., May 2002

[52] 3GPP TS 23.203, "Policy and Charging Control Architecture"

[53] IETF RFC4301, "Security Architecture for the Internet Protocol," S. Kent et al., December 2005

[54] IETF RFC4303, "IP Encapsulating Security Payload (ESP)," S. Kent, December 2005

[55] IETF RFC3948, "UDP Encapsulation of IPsec ESP Packets," A. Huttunen, January 2005

[56] IETF RFC3711, "The Secure Real-Time Transport Protocol (SRTP)," M. Baugher, March 2004

[57] UK NICC (Network Interoperability Consultative Committee) document ND1611, "Multi-Service NGN Interconnect Common Transport (issue 1)," May 2006

[58] ITU-T Recommendation Q.721, "Signalling System No. 7 – Functional Description of the Signalling System No. 7 Telephone User Part (TUP)," 1988

[59] ITU-T Recommendation Q.722, "General Function of Telephone Messages and Signals," 1988

[60] ITU-T Recommendation Q.723, "Formats and Codes," 1988

[61] ITU-T Recommendation Q.724, "Specifications of Signalling System No. 7 – Telephone User Part," 1988

[62] ITU-T Recommendation Q.725, "Signalling System No. 7 – Signalling Performance in the Telephone Application," March 1993

[63] 3GPP TS 26.071, "AMR Speech Codec; General Description"

[64] 3GPP TS 26.171, "Speech Codec Speech Processing Functions; Adaptive Multi-Rate – Wideband (AMR-WB) Speech Codec; General Description"

[65] 3GPP2 C.S0014-A v1.0, "Enhanced Variable Rate Codec, Speech Service Option 3 for Wideband Spread Spectrum Digital Systems"

[66] 3GPP TS 23.228, "IP Multimedia Subsystem (IMS); Stage 2"

[67] 3GPP TS 24.228, "Signalling Flows for the IP Multimedia Call Control based on Session Initiation Protocol (SIP) and Session Description Protocol (SDP); Stage 3"

[68] G. Camarillo, M. A. García-Martín. *The 3G IP Multimedia Subsystem (IMS): Merging the Internet and the Cellular Worlds,* John Wiley & Sons, 2006

[69] M. Poikselka, A. Niemi, H. Khartabil, G. Mayer. *The IMS: IP Multimedia Concepts and Services,* John Wiley & Sons, 2006

[70] IETF RFC4282, "The Network Access Identifier," B. Aboba et al., December 2005

[71] IETF RFS3455, "Private Header (P-Header) Extensions to the Session Initiation Protocol (SIP) for the 3rd-Generation Partnership Project (3GPP)," M. A. Garcia Martin et al., January 2003

[72] 3GPP TS 24.229, "Internet Protocol (IP) Multimedia Call Control Protocol based on Session Initiation Protocol (SIP) and Session Description Protocol (SDP); Stage 3"

[73] IETF RFC3329, "Security Mechanism Agreement for the Session Initiation Protocol (SIP)," J. Arkko, January 2003

[74] ITU-T Recommendation E.212, "The International Identification Plan for Mobile Terminals and Mobile Users," May 2004

[75] N.S0009 / IS-751, "IMSI (International Mobile Subscriber Identity) Support," February 1998

[76] 3GPP TS 23.206, "Voice Call Continuity (VCC) Between Circuit Switched (CS) and IP Multimedia Subsystem (IMS); Stage 2"

[77] 3GPP TS 24.206, "Voice Call Continuity (VCC) Between Circuit Switched (CS) and IP Multimedia Subsystem (IMS); Stage 3"

[78] TIA IS-856, "cdma2000 High Rate Packet Data Air Interface Specification," December 2001

[79] 3GPP2 X.P0042, "Voice Call Continuity"

[80] 3GPP TS 43.318, "Generic Access to the A/Gb Interface; Stage 2"

[81] 3GPP TS 44.318, "Generic Access (GA) to the A/Gb Interface; Mobile GA interface layer 3 specification"

[82] 3GPP TS 23.234, "3GPP System to Wireless Local Area Network (WLAN) Interworking; System description"

[83] IETF RFC4306, "Internet Key Exchange (IKEv2) Protocol," C. Kaufman et al., December 2005

[84] IETF RFC4186, "Extensible Authentication Protocol Method for Global System for Mobile Communications (GSM) Subscriber Identity Modules (EAP-SIM)," H. Haverinen et al., January 2006

[85] IETF RFC4187, "Extensible Authentication Protocol Method for 3rd Generation Authentication and Key Agreement (EAP-AKA)," J. Arkko et al., January 2006

[86] 3GPP TS 25.426, "UTRAN Iur and Iub Interface Data Transport & Transport Signalling for DCH Data Streams"

[87] 3GPP TS 25.427, "UTRAN Iur/Iub Interface User Plane Protocol for DCH Data Streams"

[88] 3GPP TS 25.430, "UTRAN Iub Interface: General Aspects and Principles"

[89] 3GPP TS 25.431, "UTRAN Iub Interface Layer 1"

[90] 3GPP TS 25.432, "UTRAN Iub Interface: Signalling Transport"

[91] 3GPP TS 25.433, "UTRAN Iub Interface Node B Application Part (NBAP) Signalling"

[92] 3GPP TS 25.434, "UTRAN Iub Interface Data Transport and Transport Signalling for Common Transport Channel Data Streams"

[93] 3GPP TS 25.435, "UTRAN Iub Interface User Plane Protocols for Common Transport Channel Data Streams"

[94] 3GPP TS 43.055, "Dual Transfer Mode (DTM); Stage 2"

[95] 3GPP2 A.S0009-A v2.0, "Interoperability Specification (IOS) for High Rate Packet Data (HRPD) Radio Access Network Interfaces with Session Control in the Packet Control Function"

[96] TIA/EIA/IS-136 Series, "TDMA Cellular Packet Data," September 1997

[97] TIA/EIA/IS-95, "Mobile Station – Base Station Compatibility Standard for Dual-Mode Wideband Spread Spectrum Cellular System," May 1992

[98] TIA/EIA/IS-95-A, "Mobile Station – Base Station Compatibility Standard for Dual-Mode Wideband Spread Spectrum Cellular System," May 1995

[99] TTIA-95-B, "Mobile Station – Base Station Compatibility Standard for Wideband Spread Spectrum Cellular Systems (TIA-95-B-04)(R2004)," October 2004

[100] TIA/EIA/IS-634-A, "MSC-BS Interface (A-Interface) for Public 800 Mhz (superceded by TIA/EIA-634-B)," January 1995

[101] TIA/EIA IS-41, "Cellular Radiotelecommunications Intersystem Operations"

[102] 3GPP TS 29.002, "Mobile Application Part (MAP) Specification"

[103] 3GPP TS 26.071, "AMR Speech Codec; General Description"

[104] 3GPP TS 26.171, "Speech Codec Speech Processing Functions; Adaptive Multi-Rate – Wideband (AMR-WB) Speech Codec; General Description"

[105] 3GPP TS 23.203, "Policy and Charging Control Architecture"

[106] TIA/EIA/IS-41.1-C, "Cellular Radiotelecommunications Intersystem Operations: Functional Overview (superceded by TIA/EIA-41-D)," January 1996

[107] TTIA/EIA/IS-41.3-C, "Cellular Radiotelecommunications Intersystem Operations: Automatic Roaming (superceded by TIA/EIA-41.3-D)," January 1996

[108] 3GPP2 TSG-X X.S0011-001-D v1.0, "cdma2000 Wireless IP Network Standard: Introduction," March 2006

[109] 3GPP2 TSG-X X.S0011-002-D v1.0, "cdma2000 Wireless IP Network Standard: Simple IP and Mobile IP Access Services," March 2006

[110] 3GPP2 TSG-C C.S0024-B v1.0, "cdma2000 High Rate Packet Data Air Interface Specification," June 2006

[111] 3GPP2 TSG-A A.S0009-A v2.0, "Interoperability Specification (IOS) for High Rate Packet Data (HRPD) Radio Access Network Interfaces with Session Control in the Packet Control Function"

[112] 3GPP2 TSG-S S.R0108-0 v1.0, "HRPD-cdma2000 1x Interoperability for Voice and Data System Requirements 2006/03"

[113] 3GPP2 TSG-C C.S0024-A v1.0, "cdma2000 High Rate Packet Data Air Interface Specification (5.4MB)," April 2004

[114] IEEE 802.16, "Conformance 1: Part 1: Protocol Implementation Conformance Statement (PICS) Proforma for 10-66 GHz WirelessMAN-SC Air Interface," 2003

[115] IEEE 802.16-2004, "Standard for Local and Metropolitan Area Networks Part 16: Air Interface for Fixed Broadband Wireless Access Systems," 2004

[116] IEEE 802.16e-2005 & 802.16/COR1 Part 16, "Amendment for Physical & Medium Access Control Layers for Combined Fixed & Mobile Operation," 2005

[117] IETF draft-ietf-netlmm-proxymip6-01.txt, "Proxy Mobile IPv6," S. Gundavelli et al., June 18, 2007, Work in Progress

[118] IETF draft-ietf-mip4-rfc3344bis-05, "IP Mobility Support for IPv4, Revised," C. Perkins et al., July 2007, Work in Progress

[119] IETF RFC3775, "Mobility Support in IPv6," D. Johnson et al., June 2004

[120] 3GPP TS 23.401, "General Packet Radio Service (GPRS) Enhancements for Long Term Evolution (LTE) Access"

[121] 3GPP TS 23.402, "3GPP System Architecture Evolution (SAE): Architecture Enhancements for Non-3GPP Accesses"

[122] IETF RFC4306, "Internet Key Exchange (IKEv2) Protocol," C. Kaufman, December 2005

[123] IETF RFC4186, "Extensible Authentication Protocol Method for Global System for Mobile Communications (GSM) Subscriber Identity Modules (EAP-SIM)," H. Haverinen et al., January 2006

[124] IETF RFC4187, "Extensible Authentication Protocol Method for 3rd Generation Authentication and Key Agreement (EAP-AKA)," J. Arkko et al., January 2006

[125] IETF RFC3711, "The Secure Real-time Transport Protocol (SRTP)," M. Baugher et al., March 2004

[126] IEEE 802.11-1997, "IEEE Standard for Wireless LAN Medium Access Control (MAC) and Physical Layer (PHY) specifications," 1997

[127] IEEE 802.3-1985, "Local & Metropolitan Area Networks," 1985

[128] 3GPP TS 23.153, "Out of Band Transcoder Control; Stage 2"

[129] ITU-T Recommendation Q.730, "Signalling System No.7: ISDN User Part Supplementary Services," September 1997

[130] ITU-T Recommendation Q.704, "Signalling Network Functions and Messages," July 1996

[131] IEEE 802.11a-1999, "Standard for Local and Metropolitan Area Networks – Specific Requirements Part 11: Wireless LAN Medium Access Control (MAC) and Physical Layer (PHY) Specifications Amendment 1: High-Speed Physical Layer in the 5 GHz Band," 1999

[132] IEEE 802.11b-1999, "Supplement to Standard for Local and Metropolitan Area Networks. Part 11: Wireless LAN Medium Access Control (MAC) and Physical Layer (PHY) Specifications, Higher-Speed Physical Layer (PHY) Extension in the 2.4 GHz Band," 2001

[133] IEEE 802.11g-2003, "Standard for Local and Metropolitan Area Networks: Wireless LAN Medium Access Control (MAC) and Physical Layer (PHY) Specifications," 2003

[134] IEEE 802.11h-2003, "Standard for Local and Metropolitan Area Networks: Wireless LAN Medium Access Control (MAC) and Physical Layer (PHY) Specifications, Amendment 5: Spectrum and Transmit Power Management Extensions in the 5 GHz band in Europe," 2003

[135] IEEE 802.1x-2001, "Standard for Local and Metropolitan Area Networks: Port-Based Network Access Control," 2001

[136] IEEE 802.11i-2004, "Standard for Local and Metropolitan Area Networks: Medium Access Control (MAC) Security Enhancements," 2004

[137] IEEE 802.3-1998, "Standard for Local and Metropolitan Area Networks, Carrier Sense Multiple Access with Collision Detection (CSMA-CD) Access Method and Physical Layer Specification," 1998

[138] IEEE 802.11d-1999, "Standard for Local and Metropolitan Area Networks, Amendment 3: Specification for Operation in Additional Regulatory Domains," 1999

[139] IEEE 802.11e-2005, "Standard for Local and Metropolitan Area Networks, Wireless LAN Medium Access Control (MAC) and Physical Layer (PHY) Specifications, Amendment 8: Medium Access Control (MAC) Quality of Service Enhancements," 2005

[140] IETF RFC2716, "PPP EAP TLS Authentication Protocol," B. Aboba et al., October 1999

[141] IETF RFC2779, "Instant Messaging / Presence Protocol Requirements," S. Ginozsa, September 2000

[142] IETF RFC2778, "A Model for Presence and Instant Messaging," M. Day et al., February 2000

[143] IETF RFC3920, "Extensible Messaging and Presence Protocol (XMPP): Core," P. Saint-Andre et al., September 2000

[144] IETF RFC3265, "Session Initiation Protocol (SIP) – Specific Event Notification," A. B. Roach, June 2002

[145] IETF RFC3856, "A Presence Event Package for the Session Initiation Protocol (SIP)," J. Rosenberg, August 2004

[146] IETF RFC3903, "Session Initiation Protocol (SIP) Extension for Event State Publication," A. Niemi et al., October 2004

[147] IETF RFC3863, "Presence Information Data Format (PIDF)," H. Sugano et al., August 2004

[148] IETF RFC3580, "IEEE 802.1X Remote Authentication Dial In User Service (RADIUS) Usage Guidelines," P. Congdon et al., September 2003

[149] IETF RFC3857, "A Watcher Information Event Template-Package for the Session Initiation Protocol (SIP)," J. Rosenberg, August 2004

[150] IETF RFC3858, "An Extensible Markup Language (XML) Based Format for Watcher Information," J. H. Rosenberg, August 2004

[151] IETF RFC4119 , "Presence-Based GEOPRIV Location Object Format," J. Peterson, December 2005

[152] 3GPP TS 22.141, "UMTS Presence Service; Architecture and Functional Description; Stage 1"

[153] 3GPP TS 24.141, "UMTS Presence Service Using the IP Multimedia (IM) Core Network (CN) Subsystem; Stage 3"

[154] 3GPP TS 23.141, "UMTS Presence Service; Architecture and Functional Description; Stage 2"

[155] OMA-TS-Presence_Simple-Vx, "Presence SIMPLE Specification," 2006

[156] OMA-AD-Presence_Simple-Vx, "Presence SIMPLE Specification," 2006

[157] OMA-RD-PoC-V1_0-20060609-A, "Push to Talk over Cellular Requirements," 2006

[158] OMA-AD_PoC-V1_0_1-20061128-A, "Push to Talk over Cellular (PoC) – Architecture," 2006

[159] OMA-TS-PoC_XDM-V1_0_1-20061128-A, "PoC XDM Specification," 2006

[160] IMC, "vCard, The Electronic Business Card, Version 2.1," 1996

[161] ISO/IEC 18092, 2004 "Information Technology – Telecommunications and Information Exchange Between Systems – Near Field Communication – Interface and Protocol (NFCIP-1)," 2004

[162] ISO/IEC 21481, 2004 "Information Technology – Telecommunications and Information Exchange Between Systems – Near Field Communication – Interface and Protocol (NFCIP-2)," 2004

[163] ETSI TS 102 190, 2003 "Near Field Communication (NFC) IP-1; Interface and Protocol (NFCIP-1)-V1.1.1," 2003

[164] IEEE 802.15.4-2006, "Wireless Medium Access Control (MAC) and Physical Layer (PHY) Specifications for Low-Rate Wireless Personal Area Networks (WPANs)," 2006

[165] Solomon, J. *Mobile IP,* Upper Saddle River, NJ: Prentice Hall, 1998

[166] J. Eberspacher, H. Vogel, C. Bettstetter. *GSM Switching, Services, and Protocols,* West Sussex, England: Wiley and Sons, 2001

[167] H. Kaaranen, A. Ahtiainen, L. Laitinen, S. Naghian, V. Niemi. *UMTS Networks: Architecture, Mobility and Services.* West Sussex, England: Wiley and Sons, 2001

[168] V. Garg. *Wireless Network Evolution: 2G to 3G.* Upper Saddle River. NJ: Prentice Hall, 2002

[169] Frank Ohrtman. *Voice over 802.11.* Norwood, MA: Artech House, 2004

[170] Donald J. Longueuil. *Wireless Messaging Demystified.* New York, NY: McGraw-Hill, 2003

[171] "Wireless Internet Access for Mobile Subscribers Based on the GPRS/UMTS Network," J. Park, *IEEE Communications Magazine,* Volume 40, Issue 4. (April, 2001): pp. 38–49

[172] "Global Roaming in Next-Generation Networks," Zahariadis, T. Vaxevanakis, K. Tsantilas, C. Zervos, N. Nikolaou, *IEEE Communications Magazine,* Volume 40, Issue 2. (February, 2002): pp. 145–151

[173] "An Internet Infrastructure for Cellular CDMA Networks Using Mobile IP," P. McCann, T. Hiller, *IEEE Personal Communications,* Volume 7, Issue 4. (August, 2001): pp. 26–32

[174] "Mobile IP Joins Forces with AAA," C. Perkins, *IEEE Personal Communications,* vol. 7, Issue 4 (August, 2000): pp. 59–61

[175] "Capacity of an IEEE 802.11b Wireless LAN Supporting VoIP," David P. Hole and Fouad A. Tobagi, *Proc. IEEE Int. Conference on Communications (ICC),* 2004

[176] Brad Smith. "VoIP: Friend or Foe for Cellular," *Wireless Week,* June 1, 2004

[177] Phil Solis. *Voice over Wi-Fi (VoWi-Fi),* ABI Research, 2004

[178] "Convergence, Substitution, and Project Bluephone," *Visiongain,* 2004

[179] "The Battle for the Home, Fixed Mobile Convergence," CSFB Research, 2005

[180] 3GPP TS 24.008, "Mobile Radio Interface Layer 3 Specification; Core Network Protocols; Stage 3"

[181] 3GPP TS 29.060, "General Packet Radio Service (GPRS); GPRS Tunnelling Protocol (GTP) Across the Gn and Gp Interface"

[182] 3GPP TS 23.060, "General Packet Radio Service (GPRS); Service Description; Stage 2"

[183] 3GPP2, "C.S0075-0 Interworking Specification for cdma2000 1x and High Rate Packet Data Systems"

[184] 3GPP TS 23.205, "Bearer-Independent Circuit-Switched Core Network; Stage 2"

[185] 3GPP TS 23.153, "Out of Band Transcoder Control; Stage 2"

[186] 3GPP TS 29.234, "3GPP System to Wireless Local Area Network (WLAN) Interworking; Stage 3"

[187] 3GPP TS 08.08, "Mobile-Services Switching Centre – Base Station System (MSC-BSS) Interface Layer 3 Specification"

[188] ITU-T Recommendation Y.2001, "General Overview of NGN," 2004

[189] ITU-T Recommendation Y2011, "General Principles and General Reference Model for Next Generation Networks," 2004

[190] "The Future of Mobile Voice," *Pyramid Research,* 2005

Index

www.ingramcontent.com/pod-product-compliance
Lightning Source LLC
Chambersburg PA
CBHW082004190326
41458CB00010B/3069